A VISUAL ATLAS FOR

Anatomy and Physiology

Featuring art from

David Shier, Jackie Butler, and Ricki Lewis
Hole's Human Anatomy and Physiology, 10th Edition

D1300789

McGraw Hill

Boston   Burr Ridge, IL   Dubuque, IA   Madison, WI   New York   San Francisco   St. Louis
Bangkok   Bogotá   Caracas   Kuala Lumpur   Lisbon   London   Madrid   Mexico City
Milan   Montreal   New Delhi   Santiago   Seoul   Singapore   Sydney   Taipei   Toronto

The **McGraw·Hill** Companies

A VISUAL ATLAS FOR ANATOMY AND PHYSIOLOGY

Published by McGraw-Hill Higher Education, an imprint of The McGraw-Hill Companies, Inc., 1221 Avenue of the Americas, New York, NY 10020. Copyright © The McGraw-Hill Companies, Inc., 2004. All rights reserved.

Photo credits are as follows:
Plates: 8, 11, 15, 19, 24, 25, 26, 27, 30: Courtesy of John W. Hole, Jr.
Plates 61, 62, 63, 64, 65, 66, 67, 68, 69, 70: © The McGraw-Hill Companies, Inc./Karl Rubin
Figure 11.6b: © Ed Reschke
Figures 8.3a,b: Courtesy of John W. Hole, Jr.
Figure 15.45: © Kent M. Van de Graaff

 This book is printed on recycled, acid-free paper containing 10% postconsumer waste.
RECYCLED

1 2 3 4 5 6 7 8 9 0 QPD QPD 0 9 8 7 6 5 4 3

ISBN 0-07-294643-1

www.mhhe.com

Contents

Unit 1 Skeletal System

Unit 2 Joints of the Skeletal System

Unit 3 Muscular System

Unit 4 Nervous System

Unit 5 Eye and Ear

Unit 6 Heart and Lungs

Preface

This collection of images was assembled to provide students with a comprehensive resource for studying anatomical structures and a convenient place to write notes during lecture or lab. The drawings and photographs presented in this atlas have been enlarged to the maximum size the page dimensions will allow, providing large, clear images with enhanced detail.

This atlas features full coverage of the gross anatomy of the skeletal system, the muscular system, and major joints. Also included are selected illustrations and photographs of key anatomical structures from the nervous system, as well as images depicting the anatomy of key organs and sensory structures.

In creating the *Visual Atlas for Anatomy and Physiology*, McGraw-Hill aims to meet the needs of the many different types of students enrolled in anatomy and physiology courses by providing tools that complement various learning styles. If you have any suggestions for how this product can be improved in future editions, please send your comments to the address below.

Martin J. Lange
Publisher, Life Sciences
McGraw-Hill Higher Education
2460 Kerper Boulevard
Dubuque, IA 52001

Unit 1 Skeletal System

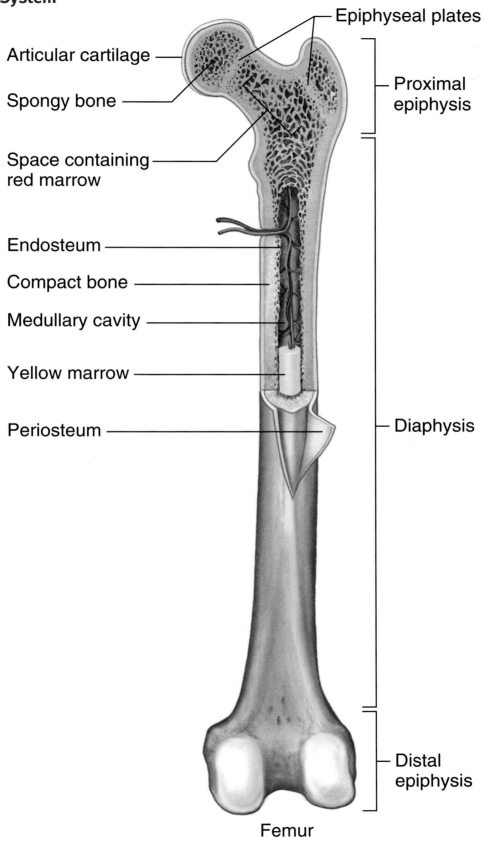

Articular cartilage

Spongy bone

Space containing red marrow

Endosteum

Compact bone

Medullary cavity

Yellow marrow

Periosteum

Epiphyseal plates

Proximal epiphysis

Diaphysis

Distal epiphysis

Femur

Figure 7.2   **Major parts of a long bone**

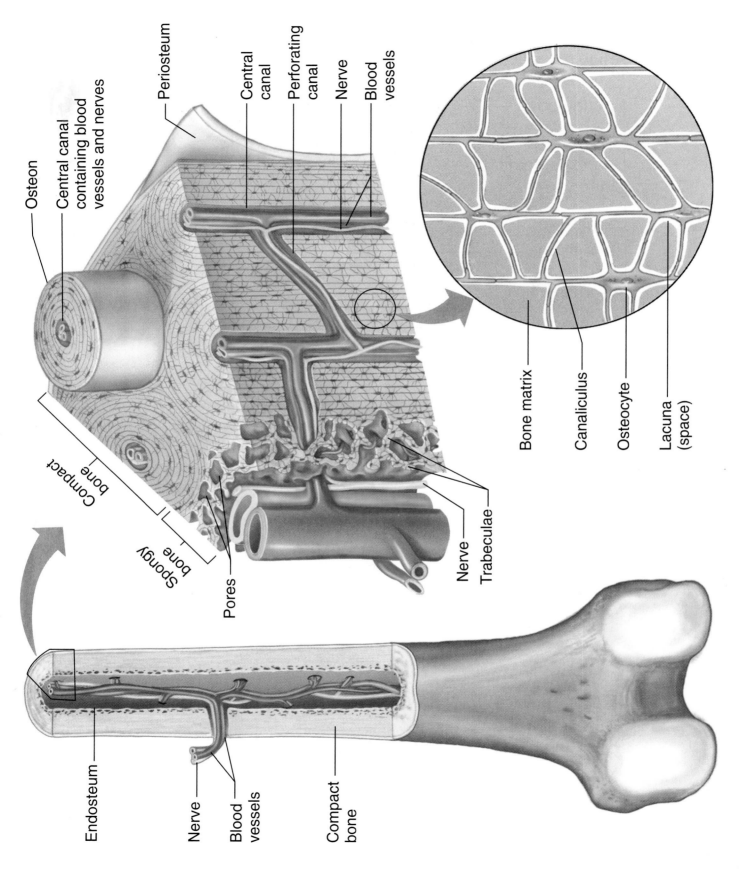

Figure 7.4   Compact bone is composed of osteons cemented together by bone matrix

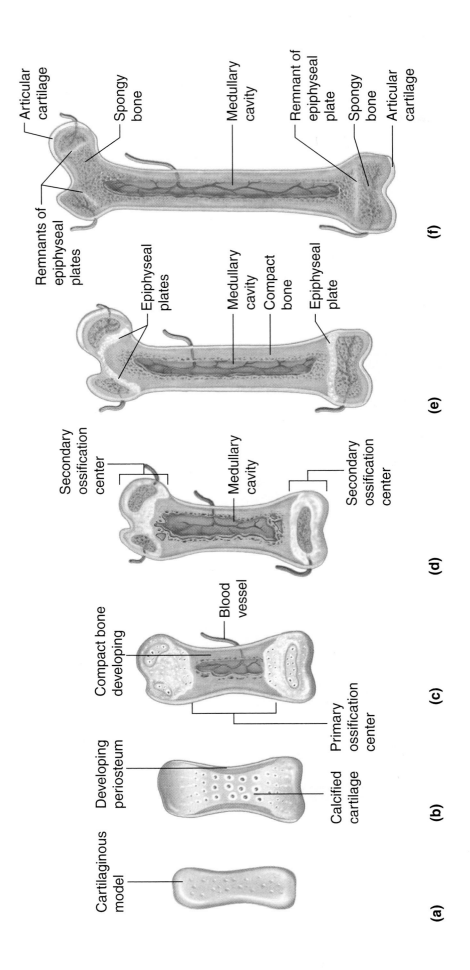

Figure 7.8   Major stages in the development of an endochondral bone

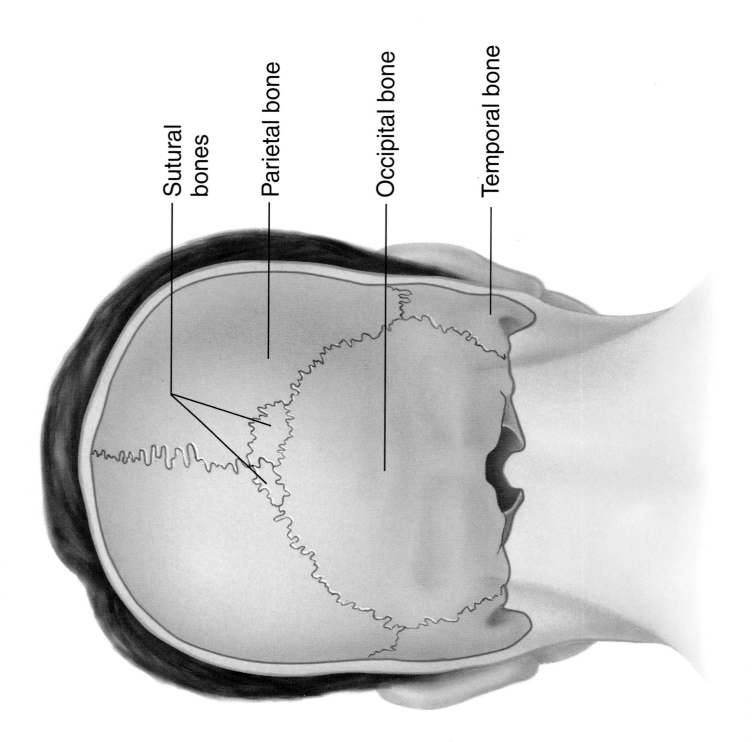

Sutural bones

Parietal bone

Occipital bone

Temporal bone

Figure 7.16   Sutural bones

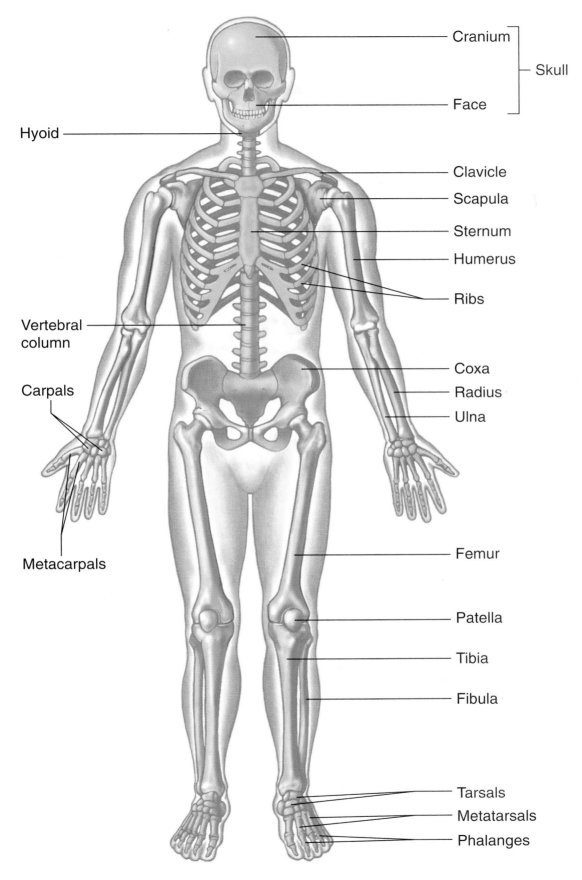

Figure 7.17a   Major bones of the skeleton (anterior view)

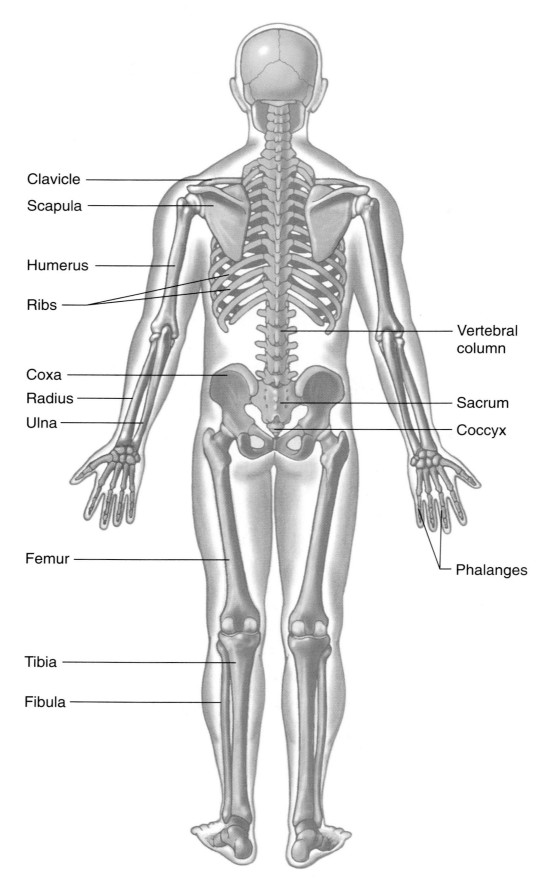

Figure 7.17b   Major bones of the skeleton (posterior view)

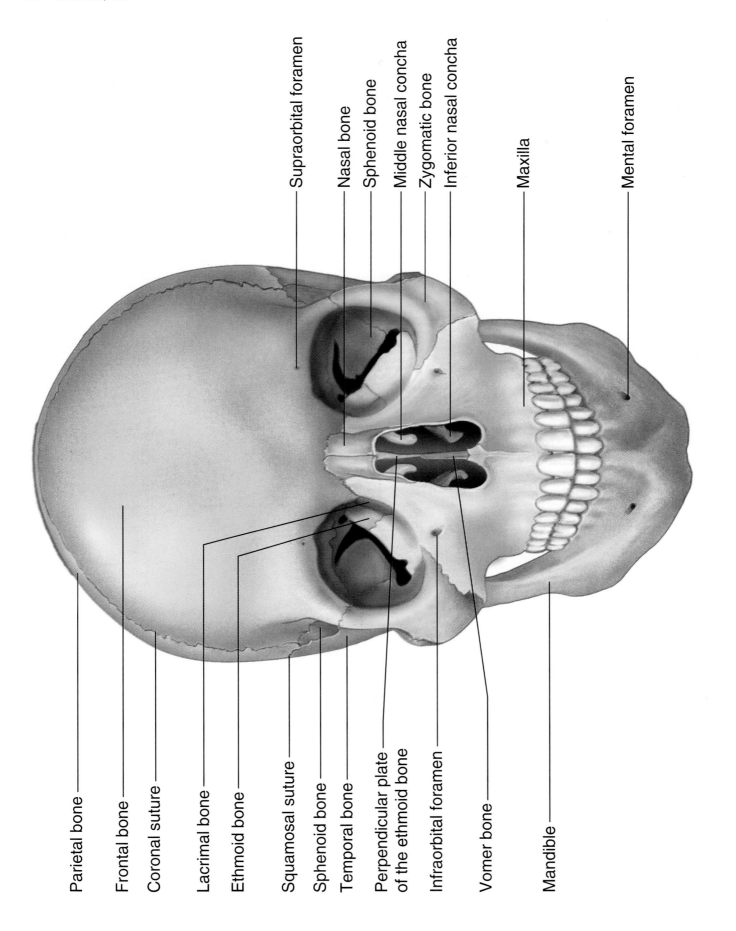

Figure 7.19 Anterior view of the skull

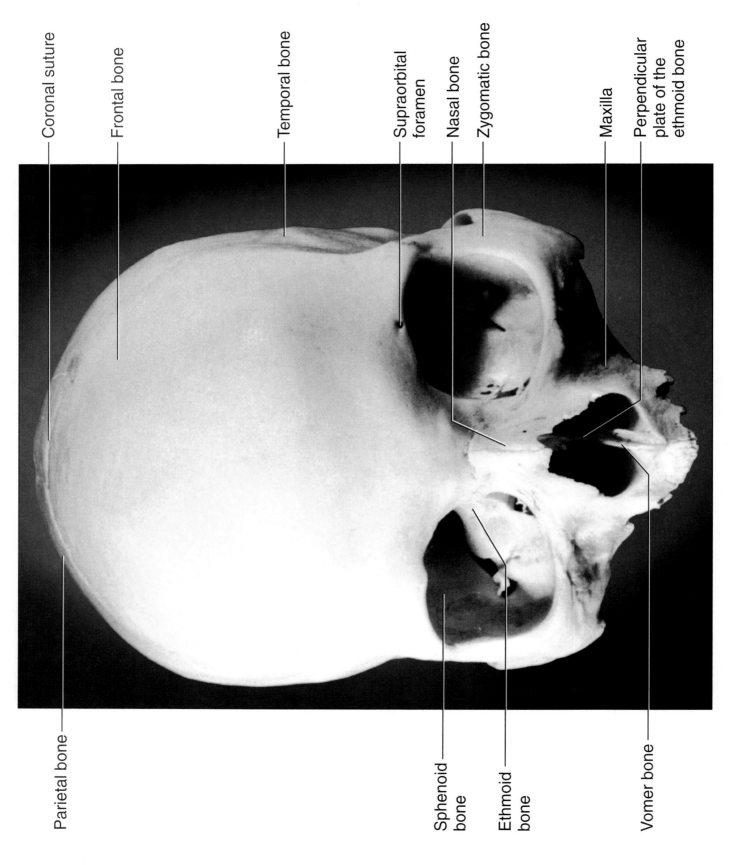

Coronal suture

Frontal bone

Temporal bone

Supraorbital foramen

Nasal bone

Zygomatic bone

Maxilla

Perpendicular plate of the ethmoid bone

Parietal bone

Sphenoid bone

Ethmoid bone

Vomer bone

Plate 8   The skull, frontal view

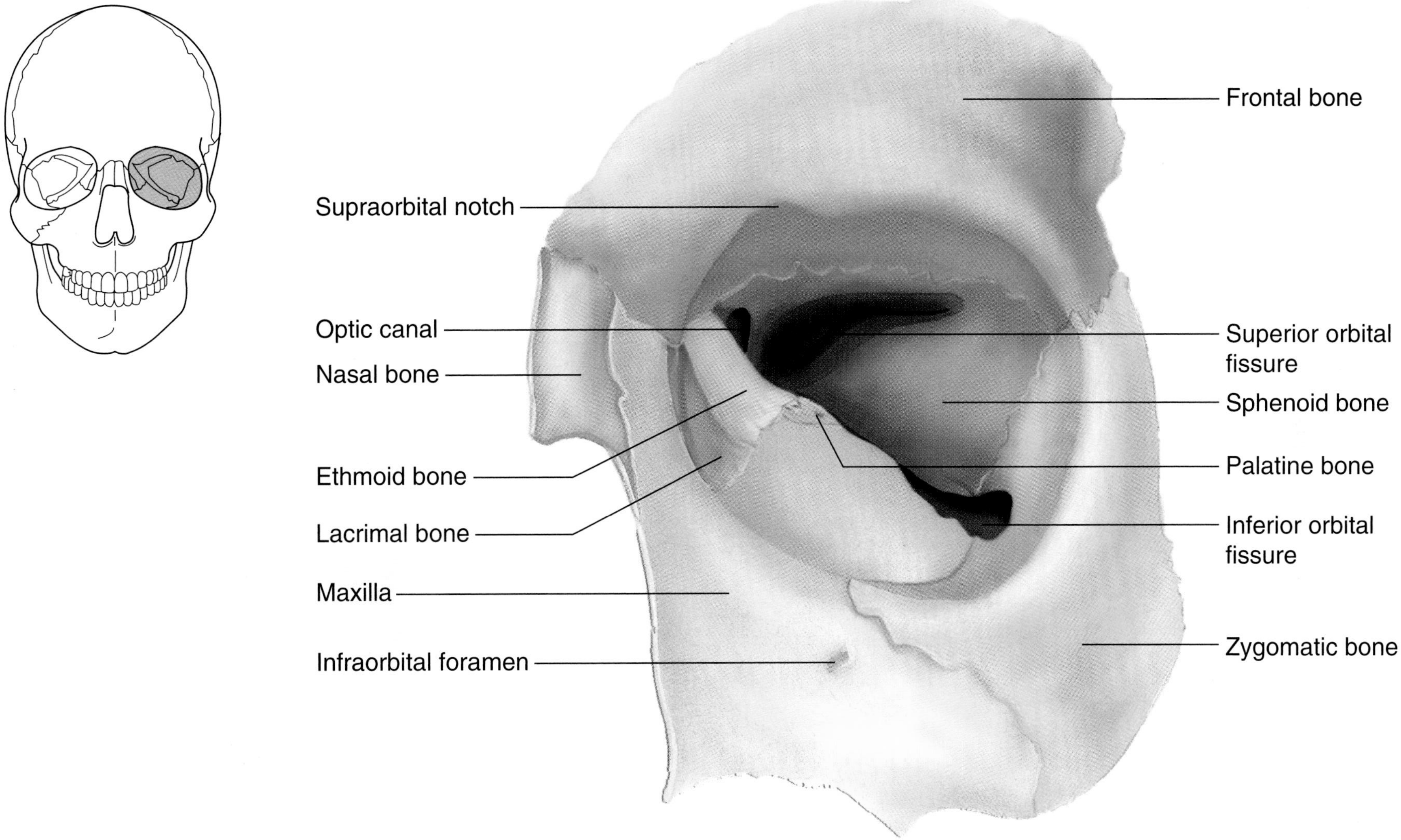

Frontal bone

Superior orbital fissure

Sphenoid bone

Palatine bone

Inferior orbital fissure

Zygomatic bone

Supraorbital notch

Optic canal

Nasal bone

Ethmoid bone

Lacrimal bone

Maxilla

Infraorbital foramen

Figure 7.20   The orbit of the eye includes both cranial and facial bones

Frontal bone

Supraorbital foramen

Zygomatic bone

Inferior orbital fissure

Nasal bone

Lacrimal bone

Ethmoid bone

Plate 11   Bones of the left orbital region

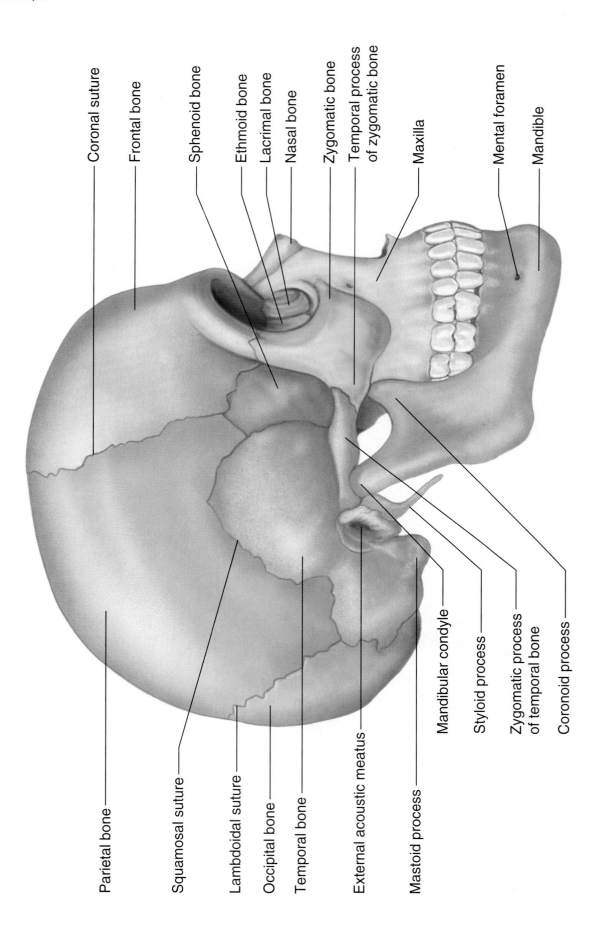

Coronal suture

Frontal bone

Sphenoid bone

Ethmoid bone

Lacrimal bone

Nasal bone

Zygomatic bone

Temporal process of zygomatic bone

Maxilla

Mental foramen

Mandible

Parietal bone

Squamosal suture

Lambdoidal suture

Occipital bone

Temporal bone

External acoustic meatus

Mastoid process

Mandibular condyle

Styloid process

Zygomatic process of temporal bone

Coronoid process

Figure 7.21   Right lateral view of the skull

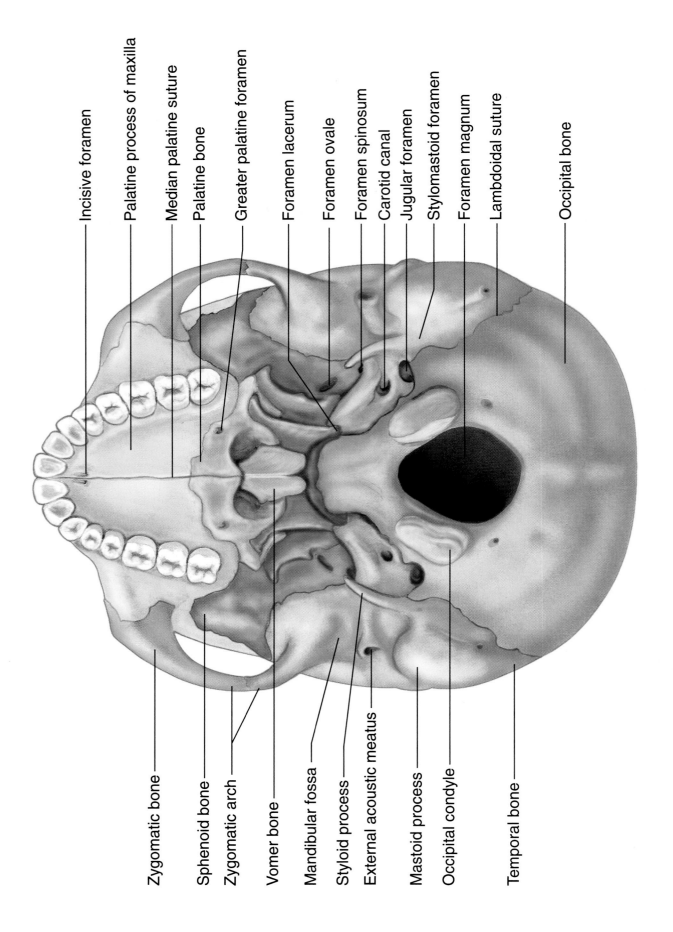

Figure 7.22   Inferior view of the skull

Incisive foramen
Palatine process of maxilla
Median palatine suture
Palatine bone
Greater palatine foramen
Foramen lacerum
Foramen ovale
Foramen spinosum
Carotid canal
Jugular foramen
Stylomastoid foramen
Foramen magnum
Lambdoidal suture
Occipital bone

Zygomatic bone
Sphenoid bone
Zygomatic arch
Vomer bone
Mandibular fossa
Styloid process
External acoustic meatus
Mastoid process
Occipital condyle
Temporal bone

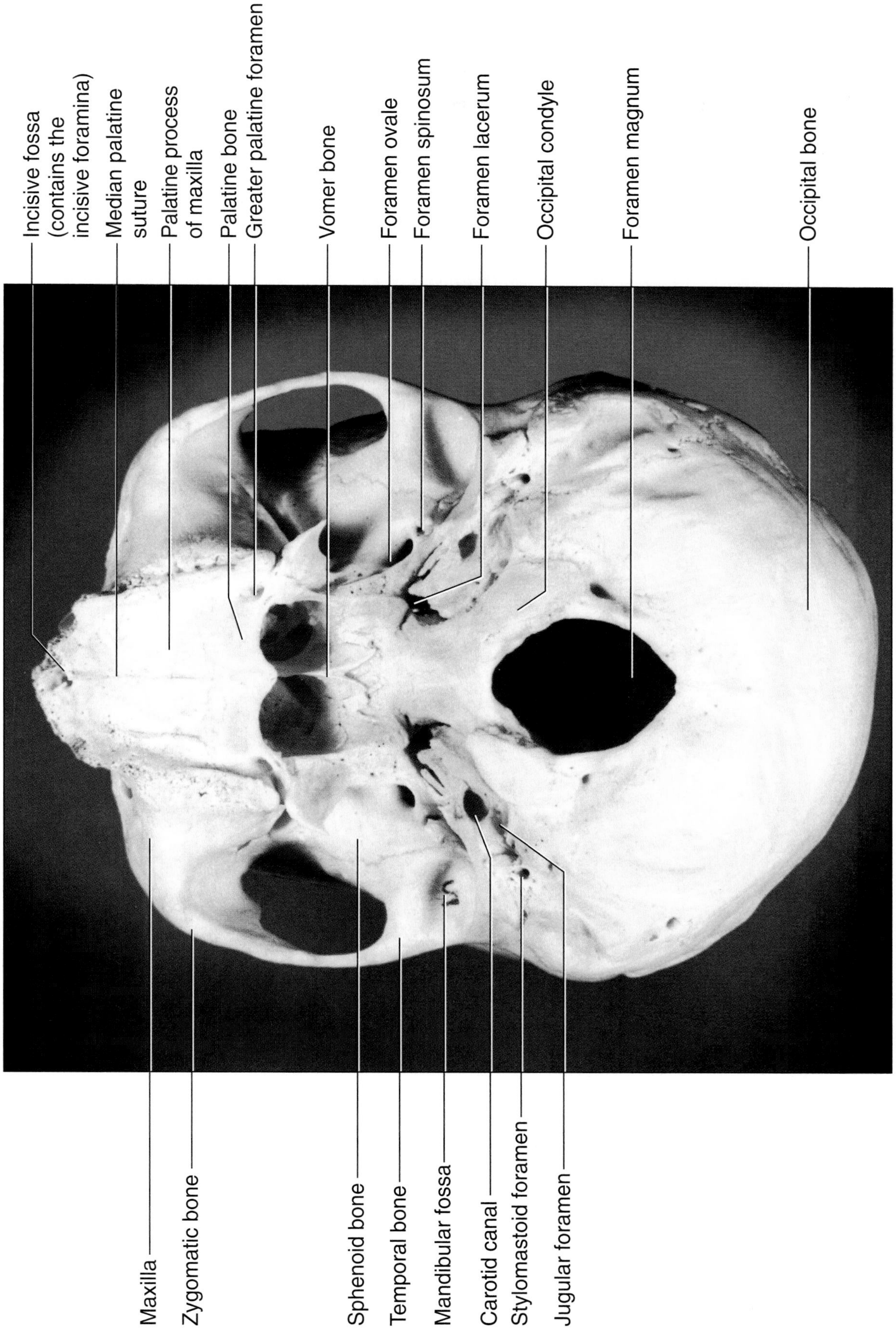

Incisive fossa (contains the incisive foramina)
Median palatine suture
Palatine process of maxilla
Palatine bone
Greater palatine foramen
Vomer bone
Foramen ovale
Foramen spinosum
Foramen lacerum
Occipital condyle
Foramen magnum
Occipital bone

Maxilla
Zygomatic bone
Sphenoid bone
Temporal bone
Mandibular fossa
Carotid canal
Stylomastoid foramen
Jugular foramen

Plate 15   The skull, inferior view

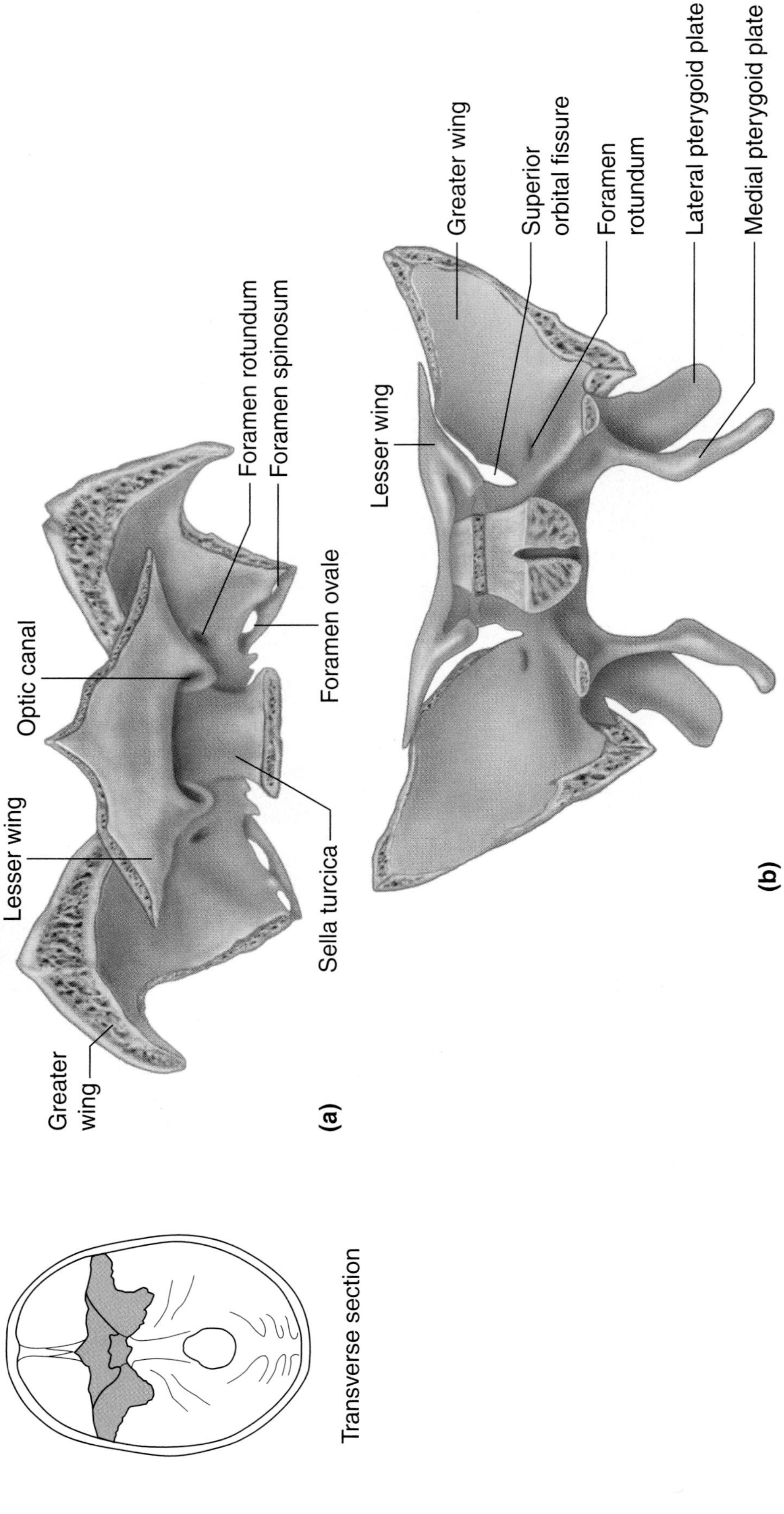

Optic canal

Foramen rotundum
Foramen spinosum

Lesser wing

Greater
wing

Foramen ovale

Sella turcica

(a)

Greater wing

Superior
orbital fissure

Foramen
rotundum

Lateral pterygoid plate

Medial pterygoid plate

Lesser wing

(b)

Transverse section

Figure 7.23   **The sphenoid bone**

Greater wing

Lesser wing

Superior orbital fissure

Foramen rotundum

Sphenoidal sinus

Plate 25   Sphenoid bone, anterior view

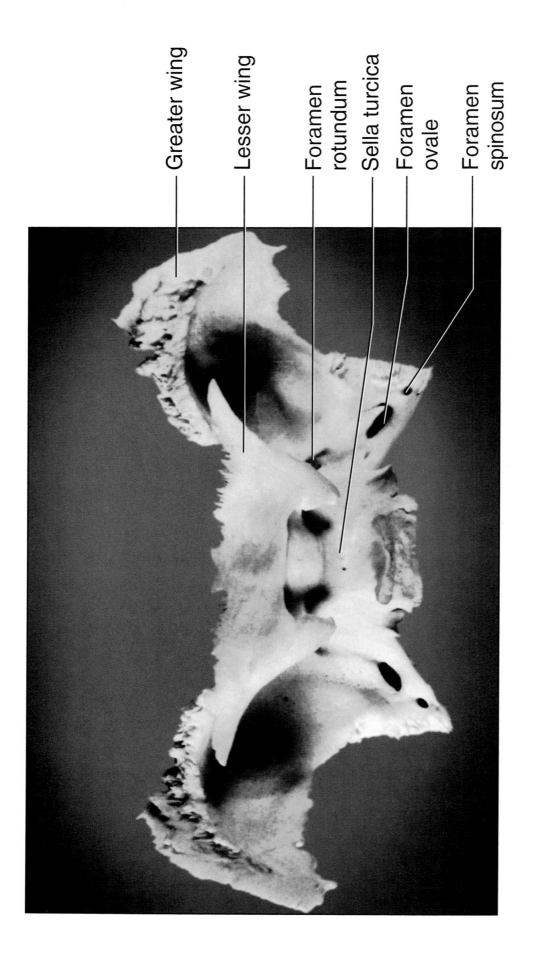

Greater wing

Lesser wing

Foramen rotundum

Sella turcica

Foramen ovale

Foramen spinosum

Plate 26   Sphenoid bone, superior view

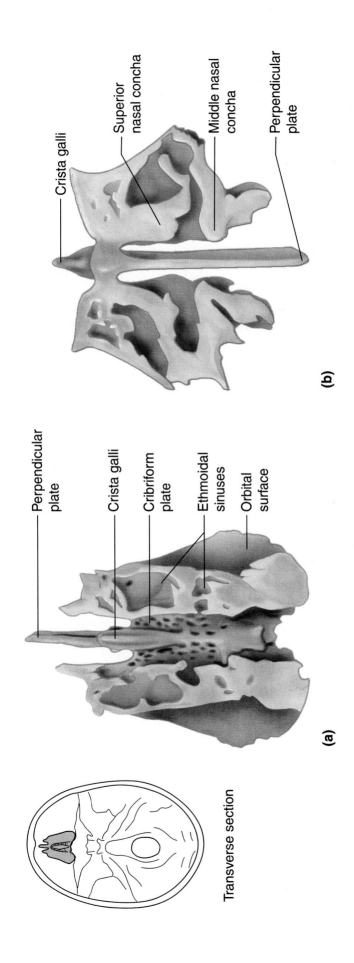

Figure 7.24   The ethmoid bone

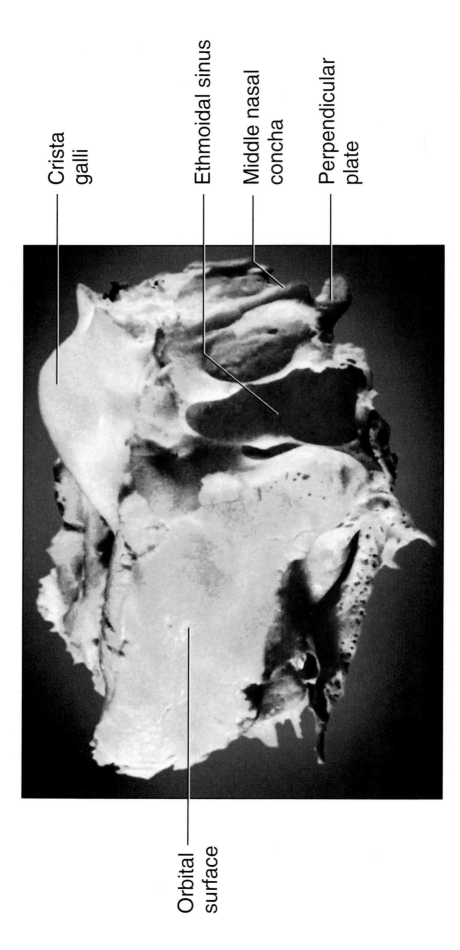

Crista galli

Ethmoidal sinus

Middle nasal concha

Perpendicular plate

Orbital surface

Plate 24   Ethmoid bone, right lateral view

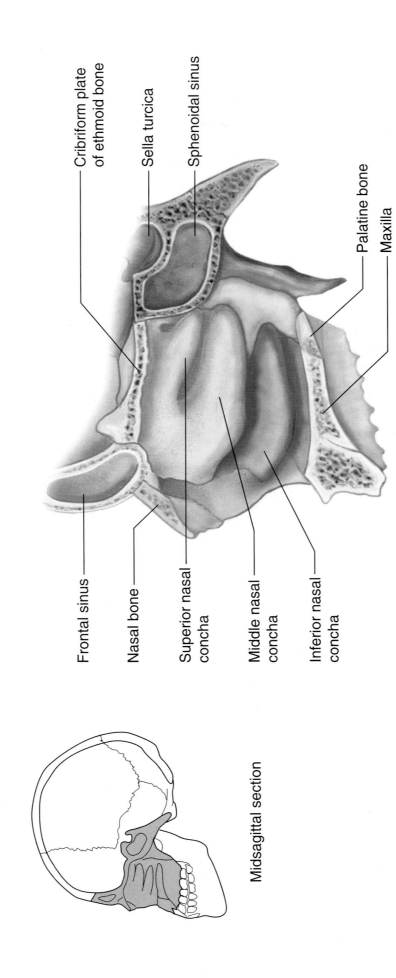

Cribriform plate of ethmoid bone

Sella turcica

Sphenoidal sinus

Palatine bone

Maxilla

Frontal sinus

Nasal bone

Superior nasal concha

Middle nasal concha

Inferior nasal concha

Midsagittal section

Figure 7.25   Lateral wall of the nasal cavity

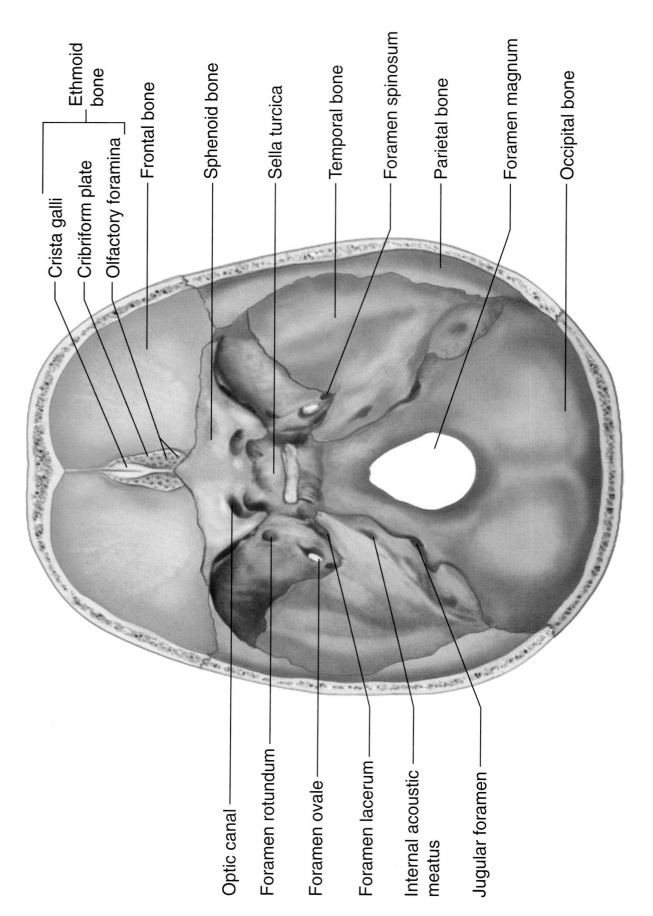

Crista galli

Cribriform plate — Ethmoid bone

Olfactory foramina

Frontal bone

Sphenoid bone

Sella turcica

Temporal bone

Foramen spinosum

Parietal bone

Foramen magnum

Occipital bone

Optic canal

Foramen rotundum

Foramen ovale

Foramen lacerum

Internal acoustic meatus

Jugular foramen

Figure 7.26 Floor of the cranial cavity

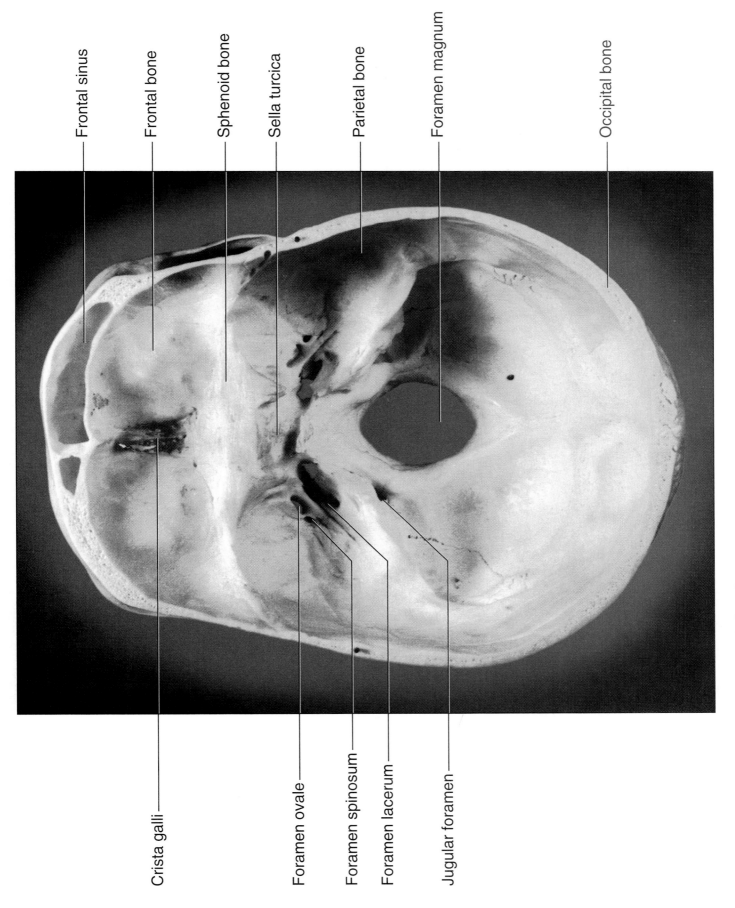

Frontal sinus

Frontal bone

Sphenoid bone

Sella turcica

Parietal bone

Foramen magnum

Occipital bone

Crista galli

Foramen ovale

Foramen spinosum

Foramen lacerum

Jugular foramen

Plate 30 The skull, floor of the cranial cavity

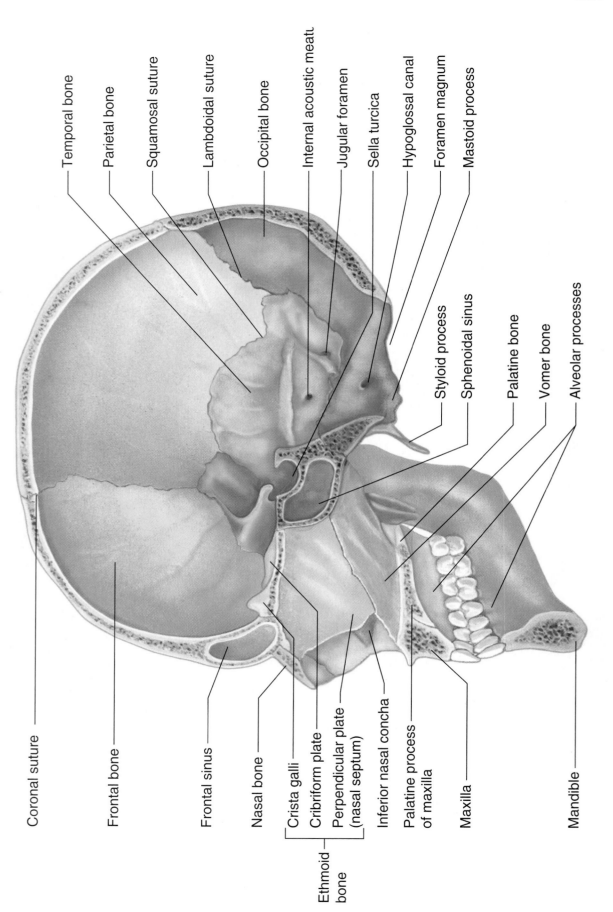

Temporal bone
Parietal bone
Squamosal suture
Lambdoidal suture
Occipital bone
Internal acoustic meatu
Jugular foramen
Sella turcica
Hypoglossal canal
Foramen magnum
Mastoid process

Styloid process
Sphenoidal sinus
Palatine bone
Vomer bone
Alveolar processes

Coronal suture
Frontal bone
Frontal sinus
Nasal bone

Ethmoid bone {
Crista galli
Cribriform plate
Perpendicular plate (nasal septum)
}

Inferior nasal concha
Palatine process of maxilla
Maxilla
Mandible

Figure 7.29 Sagittal section of the skull

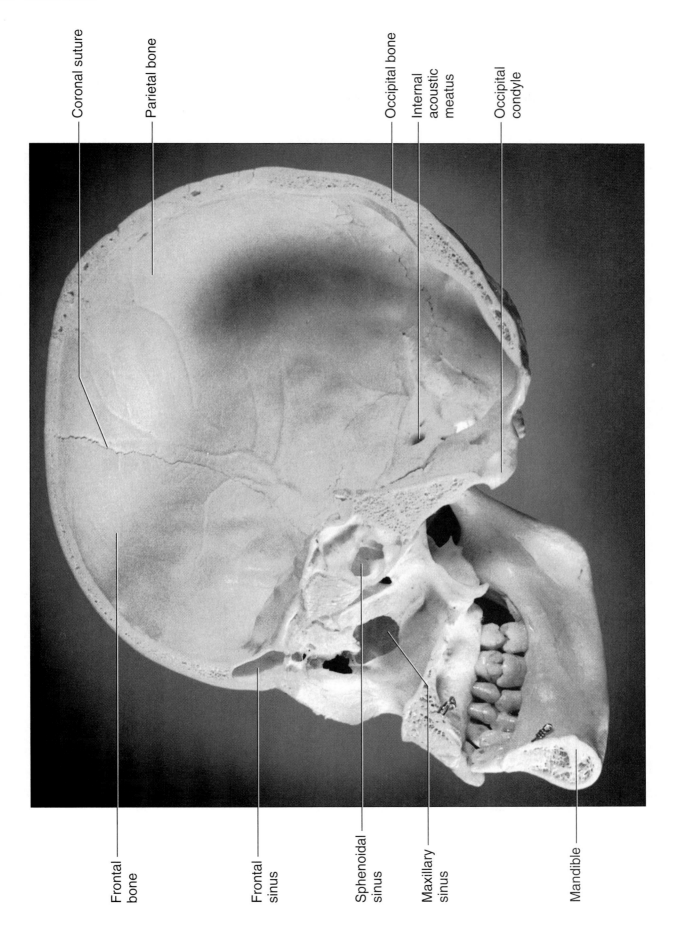

Coronal suture

Parietal bone

Occipital bone

Internal acoustic meatus

Occipital condyle

Frontal bone

Frontal sinus

Sphenoidal sinus

Maxillary sinus

Mandible

Plate 27 The skull, sagittal section

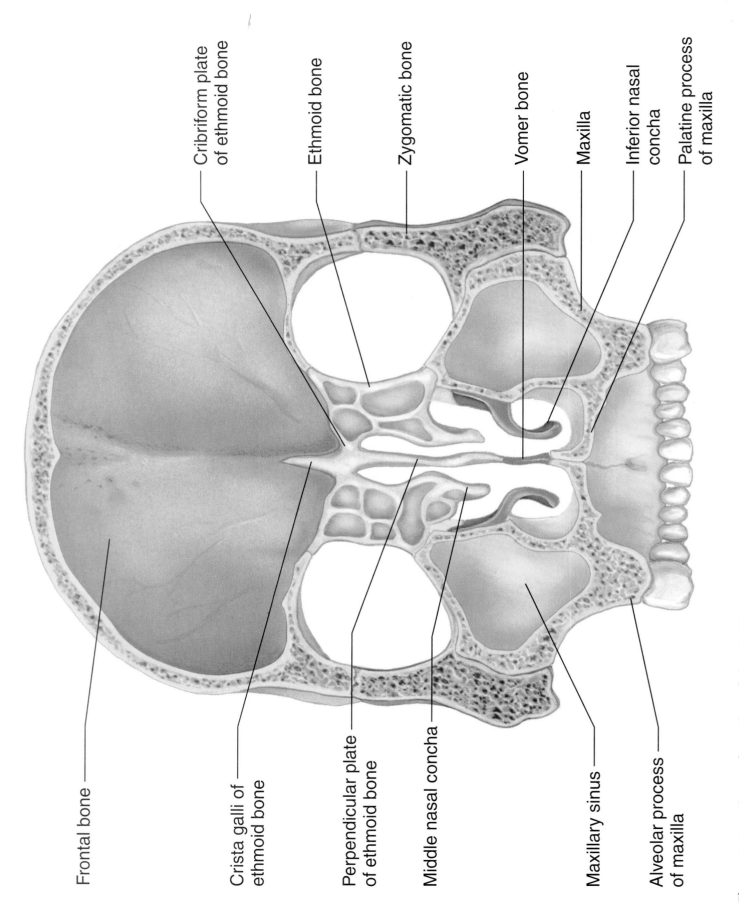

Cribriform plate
of ethmoid bone

Ethmoid bone

Zygomatic bone

Vomer bone

Maxilla

Inferior nasal
concha

Palatine process
of maxilla

Frontal bone

Crista galli of
ethmoid bone

Perpendicular plate
of ethmoid bone

Middle nasal concha

Maxillary sinus

Alveolar process
of maxilla

Figure 7.30   Coronal section of the skull

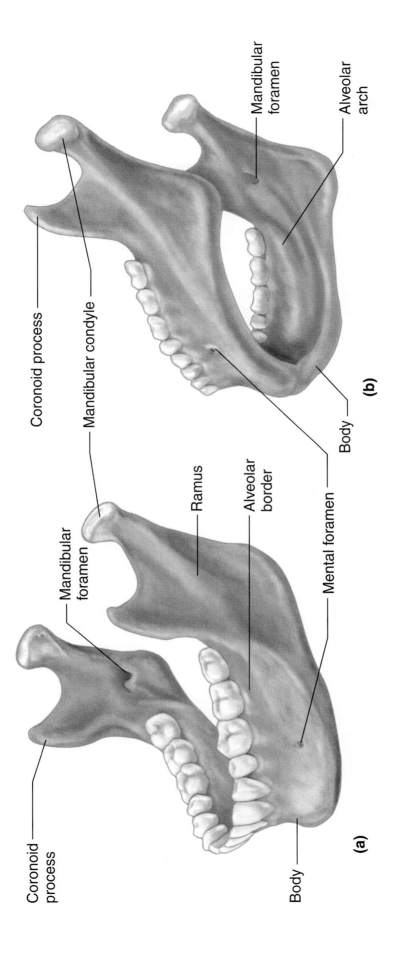

Coronoid process

Mandibular condyle

Mandibular foramen

Ramus

Alveolar border

Mental foramen

Body

Coronoid process

(a)

Mandibular foramen

Alveolar arch

Body

(b)

Figure 7.31 Mandible

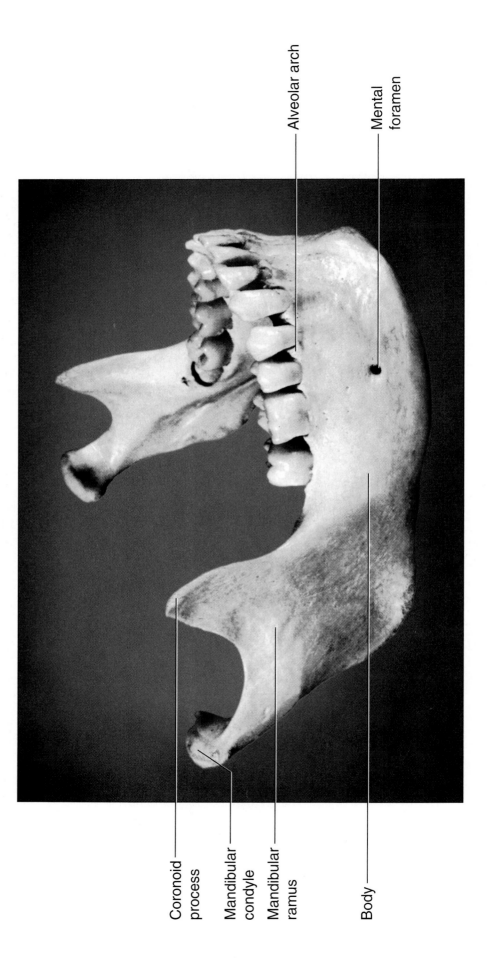

Alveolar arch

Mental foramen

Coronoid process

Mandibular condyle

Mandibular ramus

Body

Plate 19   Mandible, lateral view

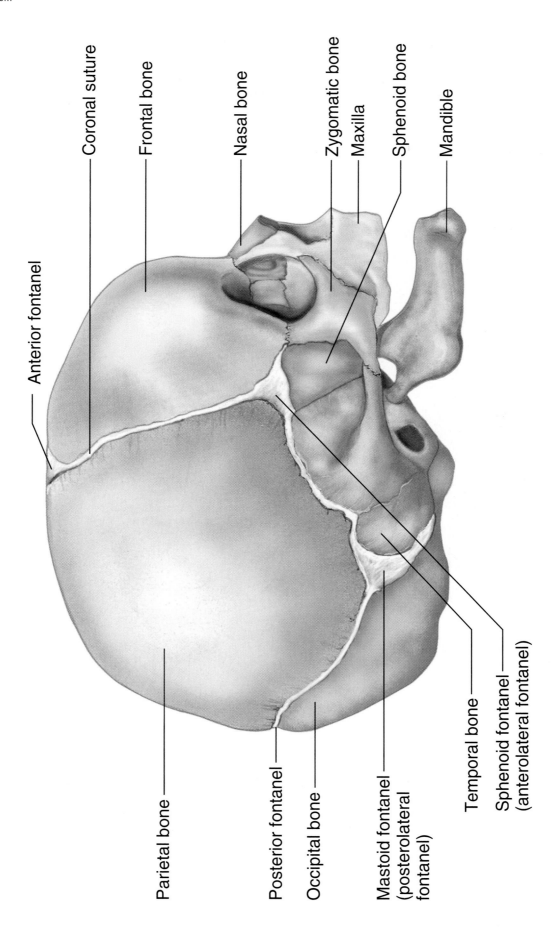

Coronal suture

Frontal bone

Nasal bone

Zygomatic bone

Maxilla

Sphenoid bone

Mandible

Anterior fontanel

Parietal bone

Posterior fontanel

Occipital bone

Mastoid fontanel (posterolateral fontanel)

Temporal bone

Sphenoid fontanel (anterolateral fontanel)

Figure 7.33a   Right lateral view of the infantile skull

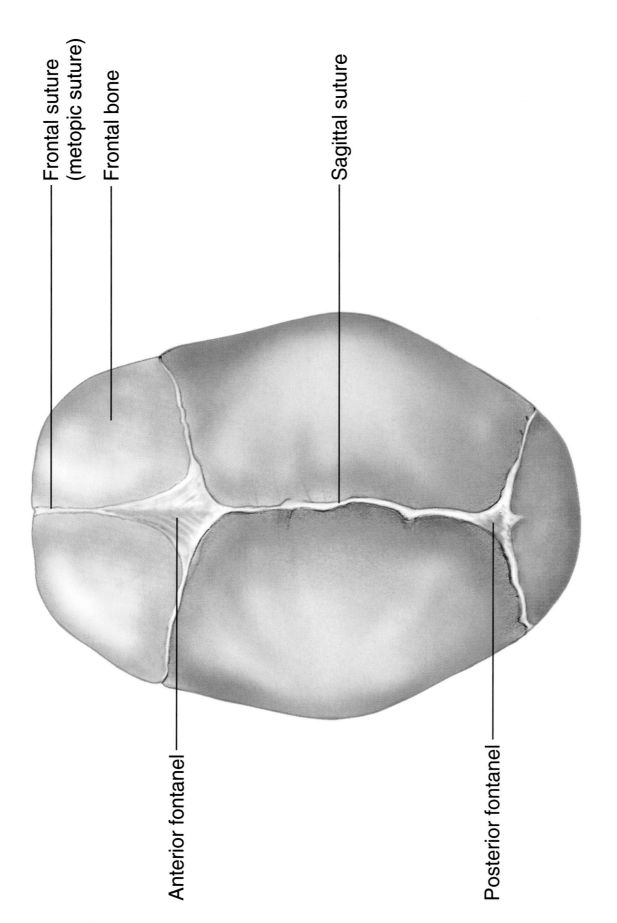

Frontal suture (metopic suture)

Frontal bone

Sagittal suture

Anterior fontanel

Posterior fontanel

Figure 7.33b   Superior view of the infantile skull

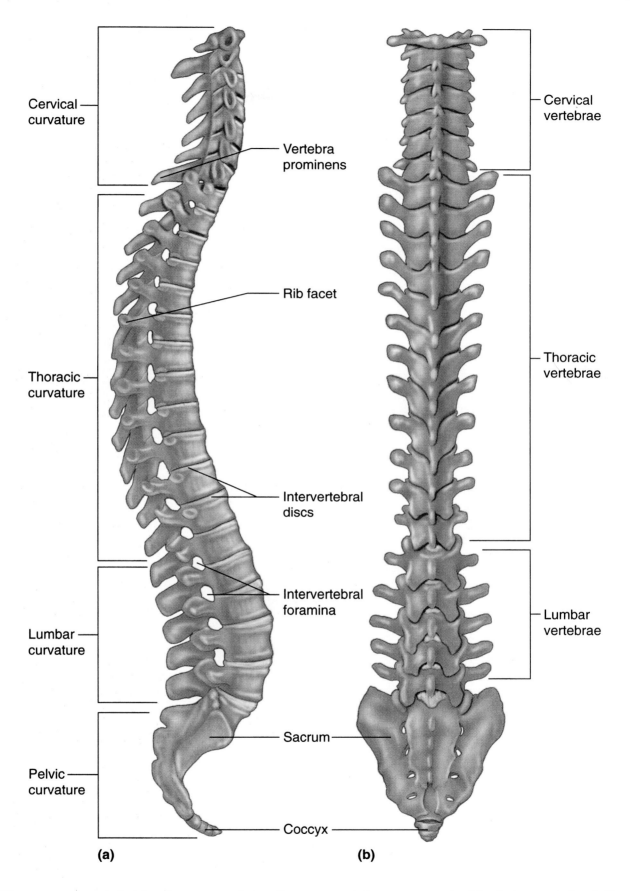

Cervical curvature

Vertebra prominens

Rib facet

Thoracic curvature

Intervertebral discs

Lumbar curvature

Intervertebral foramina

Pelvic curvature

Sacrum

Coccyx

(a)

Cervical vertebrae

Thoracic vertebrae

Lumbar vertebrae

(b)

Figure 7.34 The curved vertebral column consists of many vertebrae separated by intervertebral discs

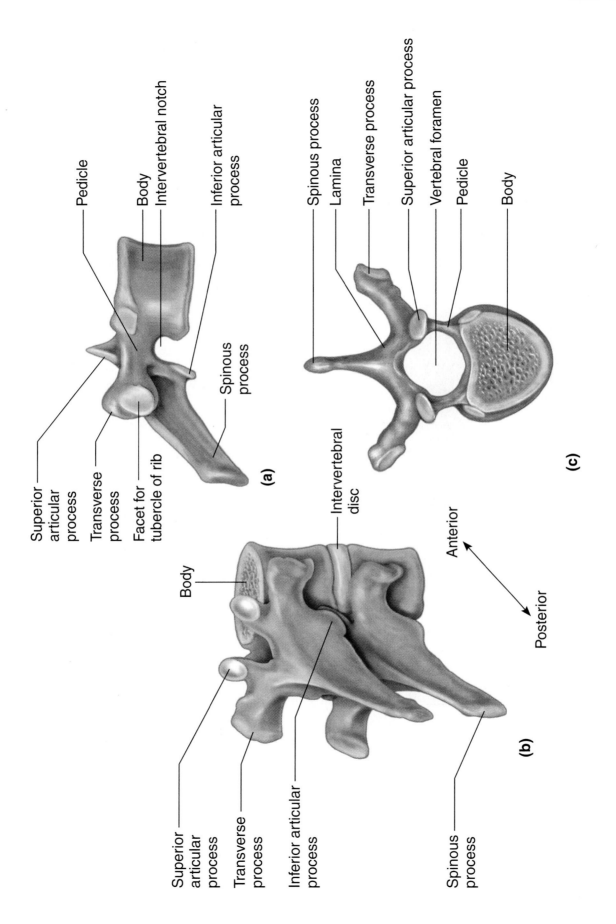

Figure 7.35   Typical thoracic vertebra

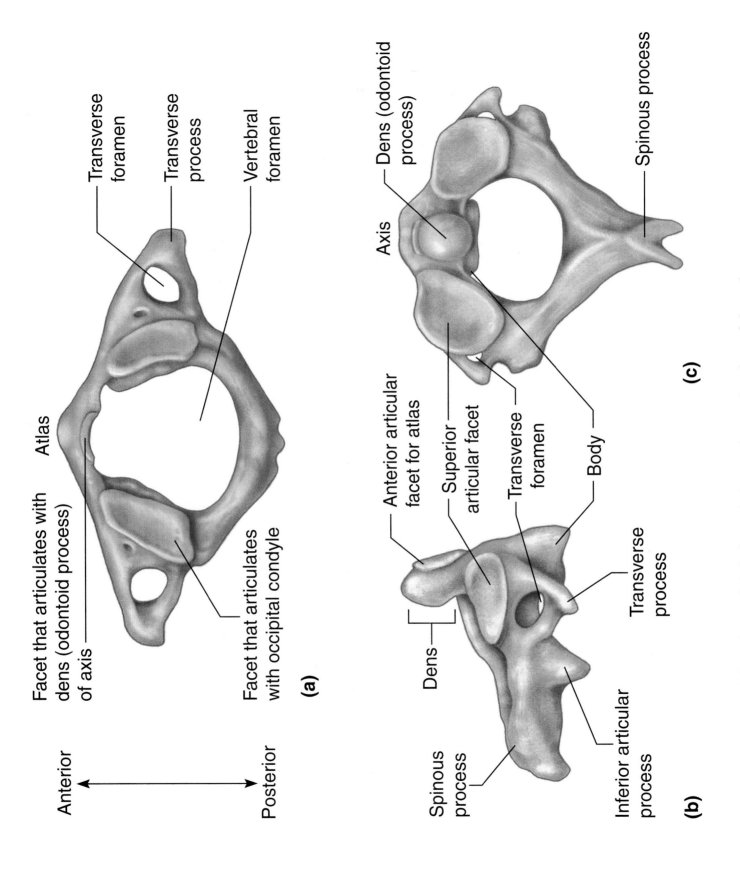

Transverse foramen

Transverse process

Vertebral foramen

Facet that articulates with dens (odontoid process) of axis

Atlas

Facet that articulates with occipital condyle

Anterior

Posterior

(a)

Dens (odontoid process)

Axis

Spinous process

Anterior articular facet for atlas

Superior articular facet

Transverse foramen

Body

Transverse process

Dens

Spinous process

Inferior articular process

(b)

(c)

Figure 7.36   Superior view of the atlas and right lateral view and superior view of the axis

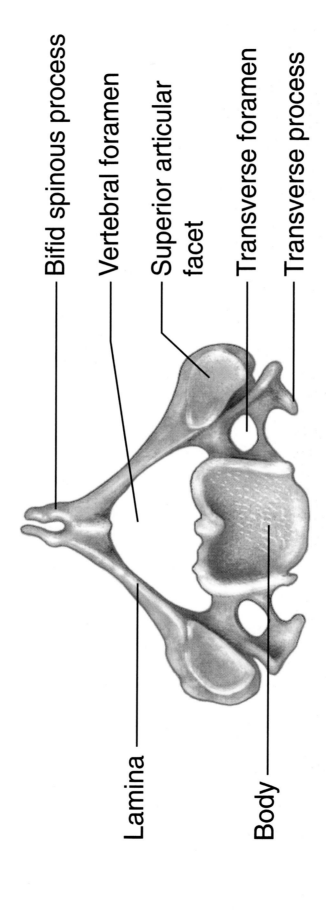

Bifid spinous process

Vertebral foramen

Superior articular facet

Transverse foramen

Transverse process

Lamina

Body

Figure 7.38a   Superior view of a cervical vertebra

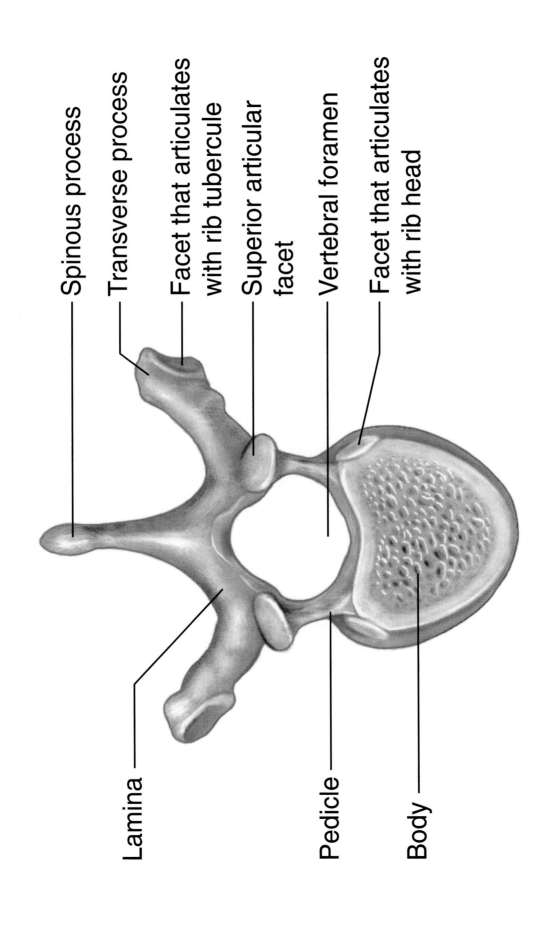

Spinous process

Transverse process

Facet that articulates with rib tubercule

Superior articular facet

Vertebral foramen

Facet that articulates with rib head

Lamina

Pedicle

Body

Figure 7.38b   Superior view of a thoracic vertebra

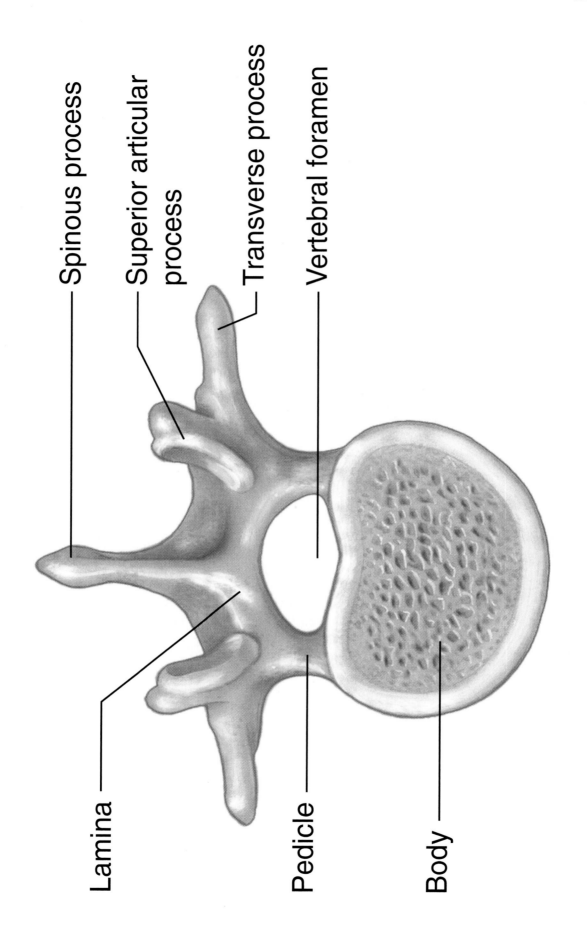

Spinous process

Superior articular process

Transverse process

Vertebral foramen

Lamina

Pedicle

Body

Figure 7.38c Superior view of a lumbar vertebra

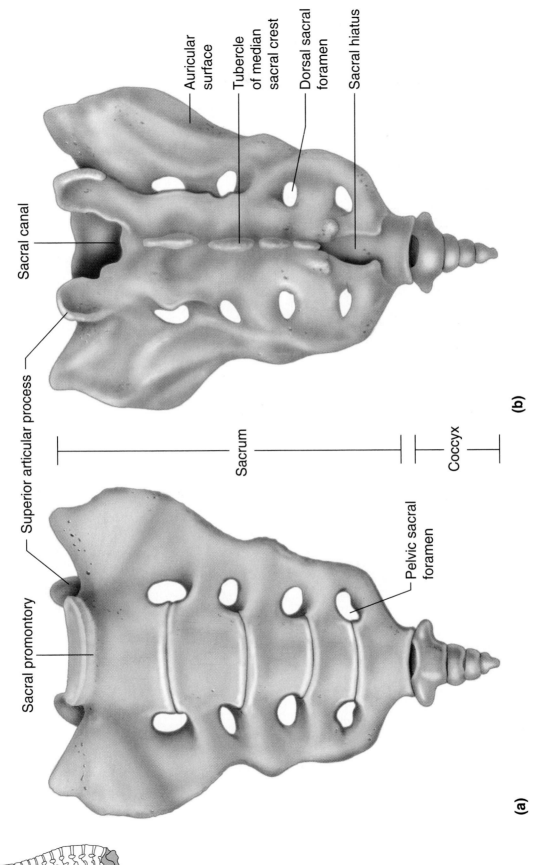

Auricular surface

Tubercle of median sacral crest

Dorsal sacral foramen

Sacral hiatus

Sacral canal

Superior articular process

Sacral promontory

Pelvic sacral foramen

Sacrum

Coccyx

(b)

(a)

Figure 7.39   Sacrum and coccyx

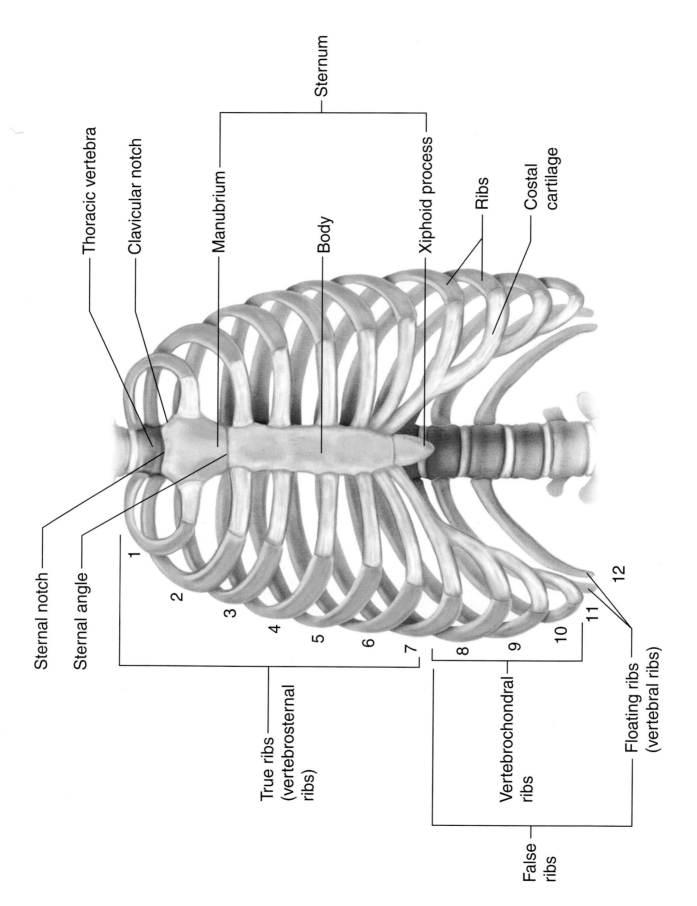

Figure 7.40a   The thoracic cage includes the thoracic vertebrae, the sternum, the ribs, and the costal cartilages

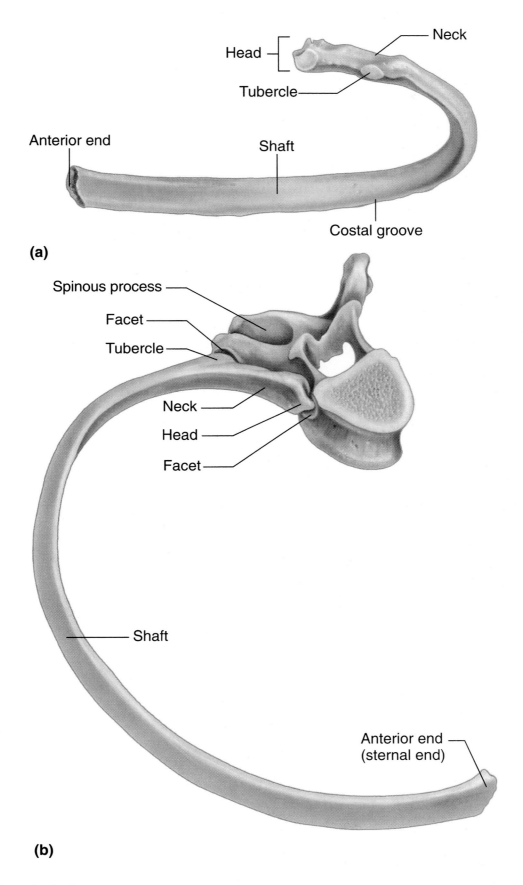

(a)

(b)

Figure 7.41   A typical rib

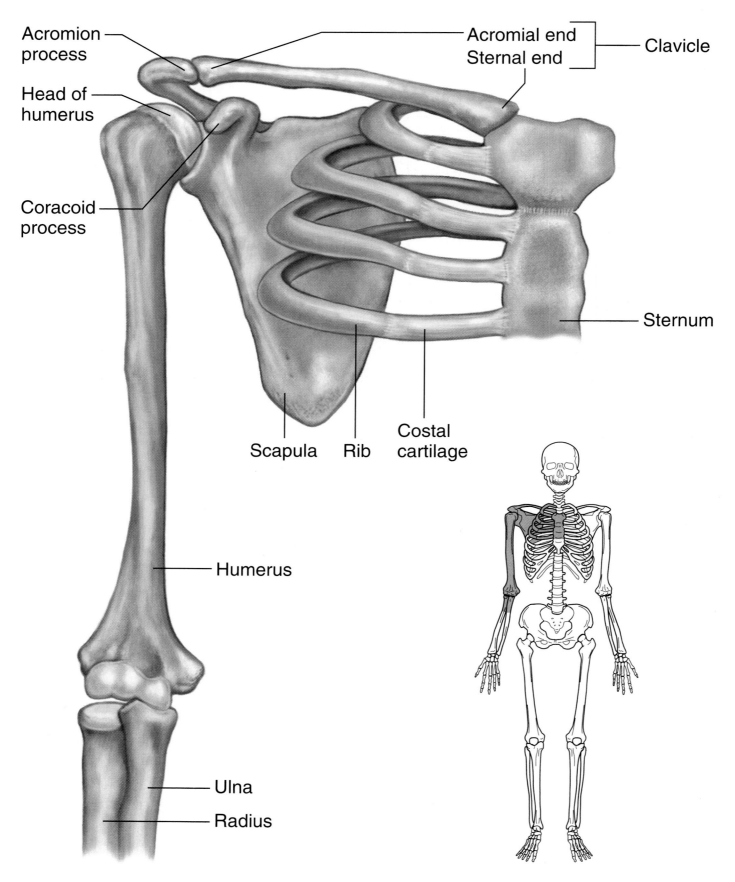

Acromion process

Head of humerus

Coracoid process

Acromial end
Sternal end

Clavicle

Sternum

Scapula   Rib   Costal cartilage

Humerus

Ulna

Radius

Figure 7.42a   **The pectoral girdle**

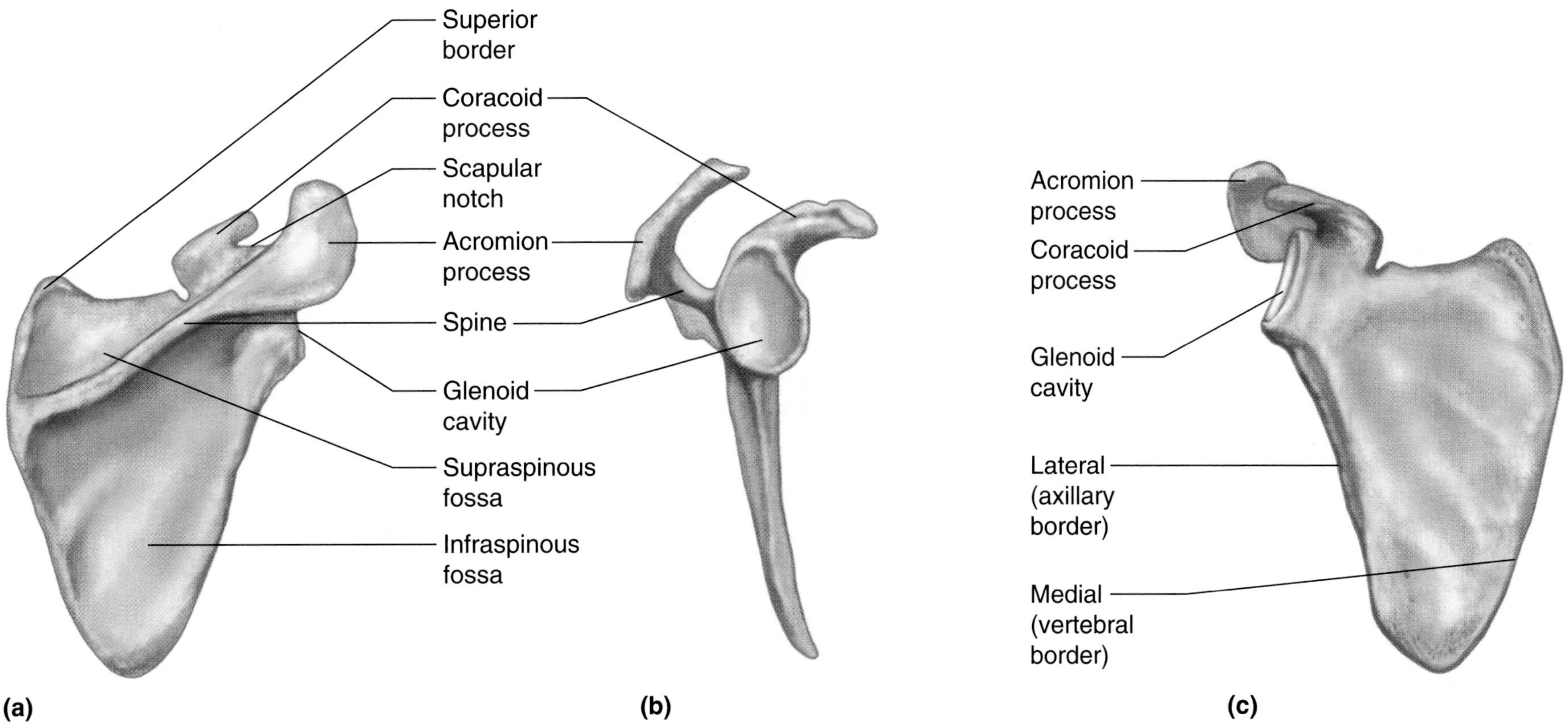

Acromion
process

Coracoid
process

Glenoid
cavity

Lateral
(axillary
border)

Medial
(vertebral
border)

(c)

Superior
border

Coracoid
process

Scapular
notch

Acromion
process

Spine

Glenoid
cavity

Supraspinous
fossa

Infraspinous
fossa

(b)

(a)

Figure 7.43 Scapula

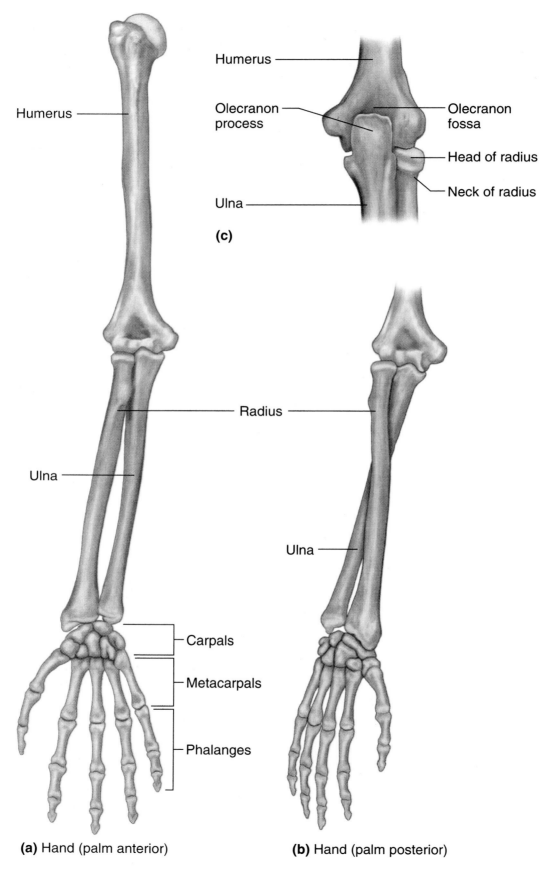

Humerus

Olecranon
process

Ulna

Olecranon
fossa

Head of radius

Neck of radius

(c)

Humerus

Radius

Ulna

Carpals

Metacarpals

Phalanges

Ulna

(a) Hand (palm anterior)

(b) Hand (palm posterior)

Figure 7.44a–c   Right upper limb

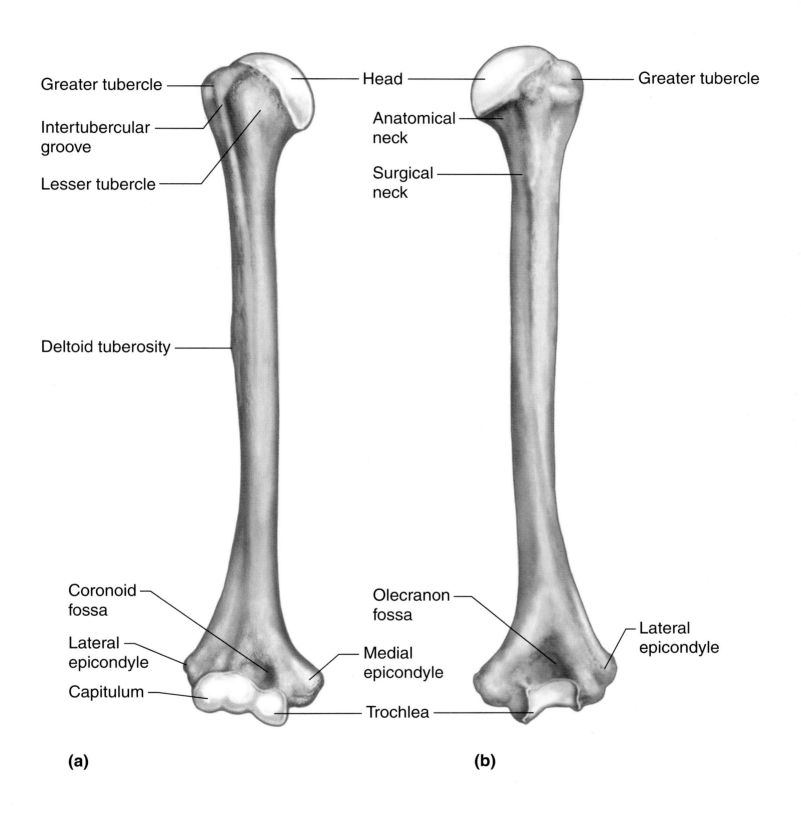

Greater tubercle

Intertubercular groove

Lesser tubercle

Deltoid tuberosity

Coronoid fossa

Lateral epicondyle

Capitulum

Head

Anatomical neck

Surgical neck

Olecranon fossa

Medial epicondyle

Trochlea

Greater tubercle

Lateral epicondyle

(a)

(b)

Figure 7.45   Humerus

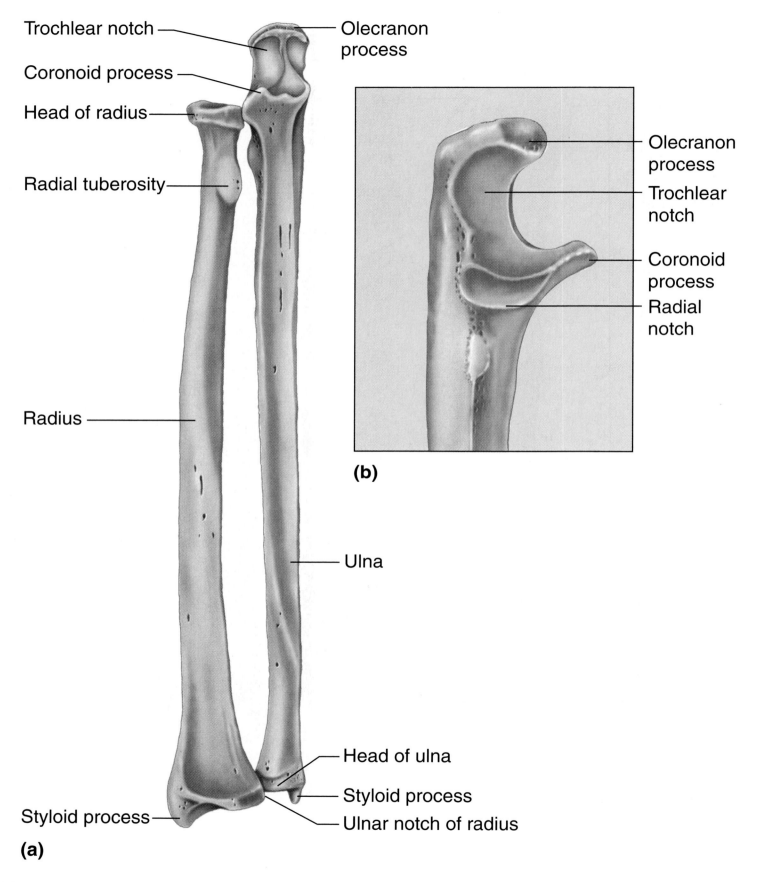

Trochlear notch

Coronoid process

Head of radius

Radial tuberosity

Radius

Styloid process

(a)

Olecranon process

Ulna

Head of ulna

Styloid process

Ulnar notch of radius

Olecranon process

Trochlear notch

Coronoid process

Radial notch

(b)

Figure 7.46   Radius and ulna

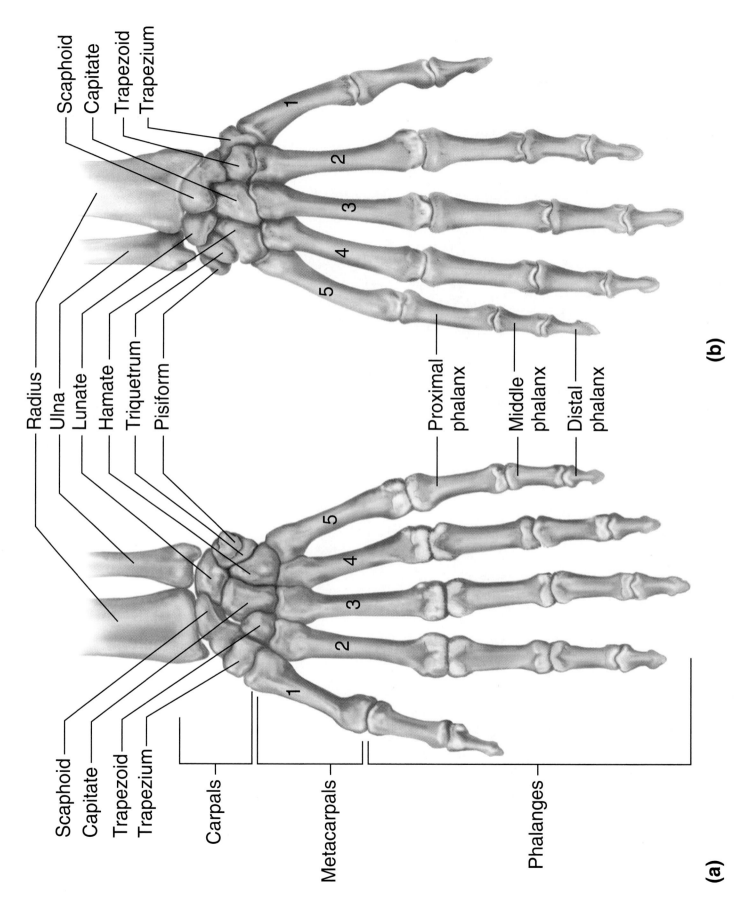

Scaphoid
Capitate
Trapezoid
Trapezium

Radius
Ulna
Lunate
Hamate
Triquetrum
Pisiform

Proximal phalanx
Middle phalanx
Distal phalanx

(b)

Scaphoid
Capitate
Trapezoid
Trapezium

Carpals

Metacarpals

Phalanges

(a)

Figure 7.47a–b Wrist and hand

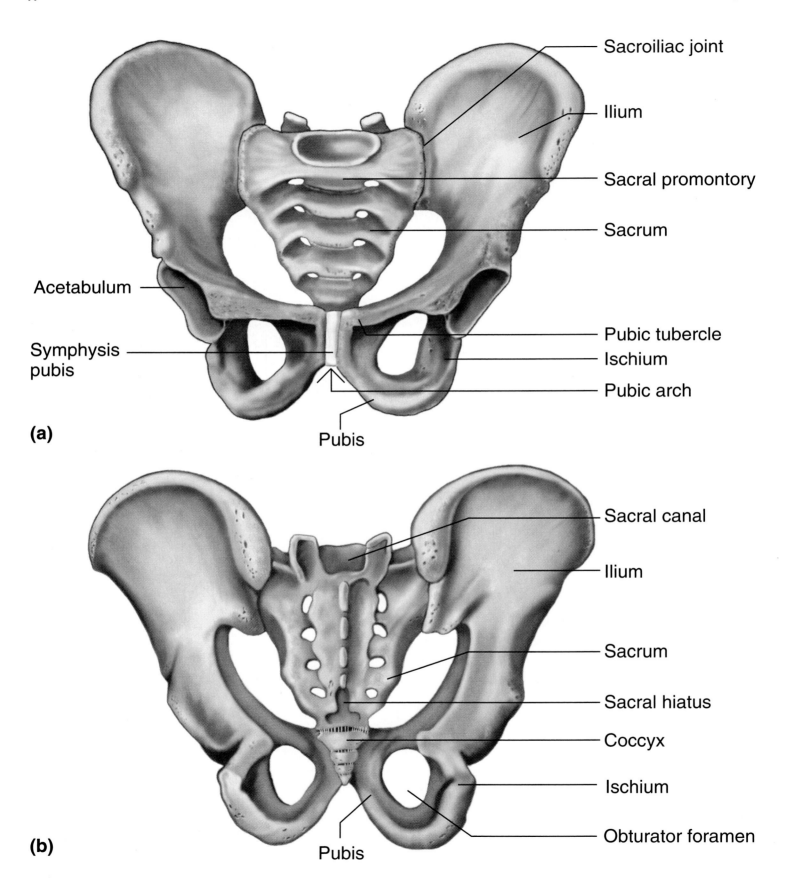

Sacroiliac joint

Ilium

Sacral promontory

Sacrum

Acetabulum

Pubic tubercle

Symphysis
pubis

Ischium

Pubic arch

(a)

Pubis

Sacral canal

Ilium

Sacrum

Sacral hiatus

Coccyx

Ischium

Obturator foramen

(b)

Pubis

Figure 7.49a–b   Pelvic girdle

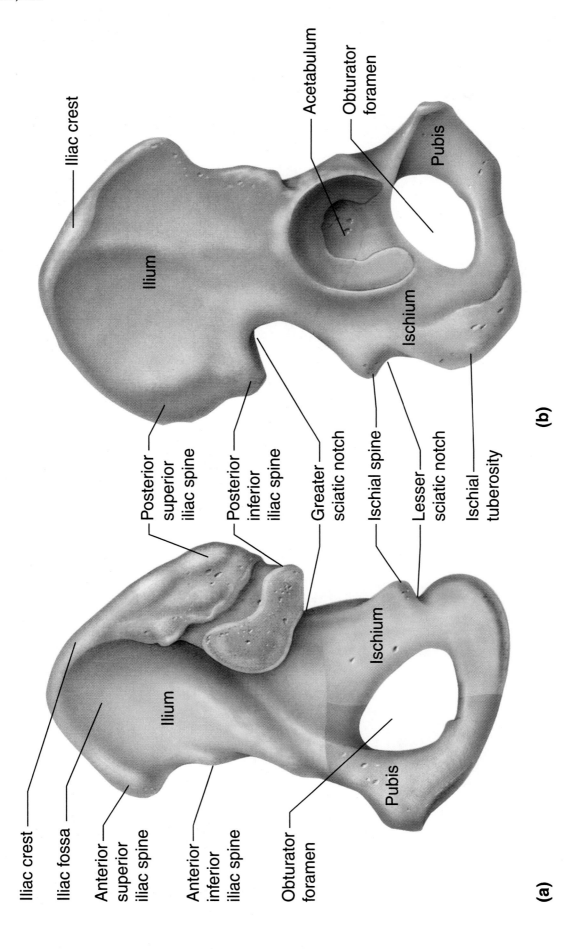

Iliac crest

Acetabulum

Obturator foramen

Pubis

Ilium

Ischium

Posterior superior iliac spine

Posterior inferior iliac spine

Greater sciatic notch

Ischial spine

Lesser sciatic notch

Ischial tuberosity

(b)

Iliac crest

Iliac fossa

Anterior superior iliac spine

Anterior inferior iliac spine

Obturator foramen

Ilium

Ischium

Pubis

(a)

Figure 7.50   **Coxa**

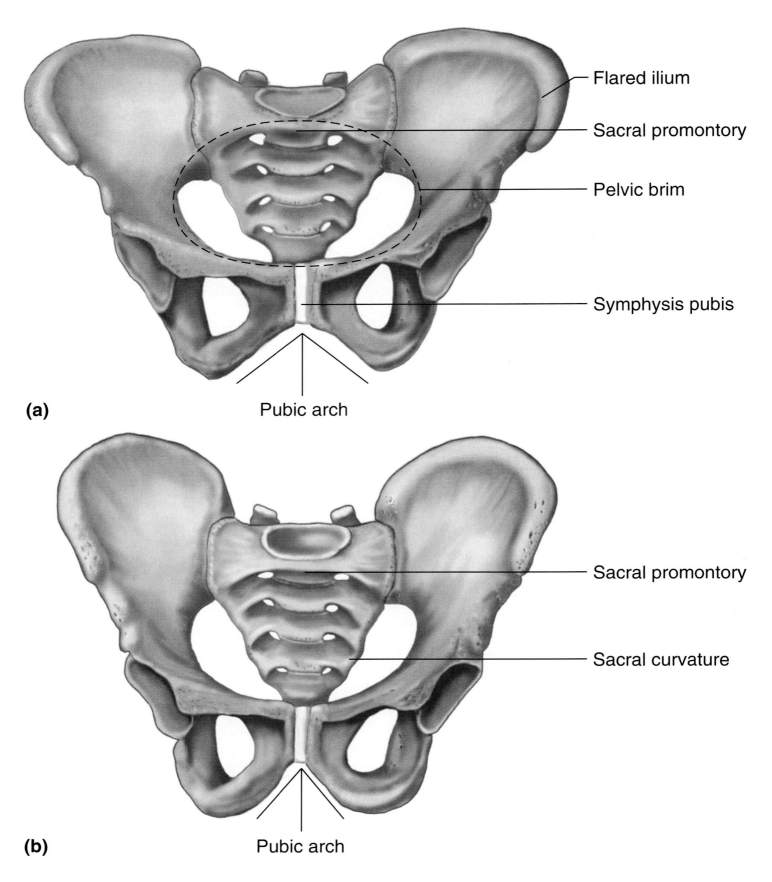

(a)

Flared ilium

Sacral promontory

Pelvic brim

Symphysis pubis

Pubic arch

(b)

Sacral promontory

Sacral curvature

Pubic arch

Figure 7.51   The female pelvis is usually wider in all diameters and roomier than that of the male

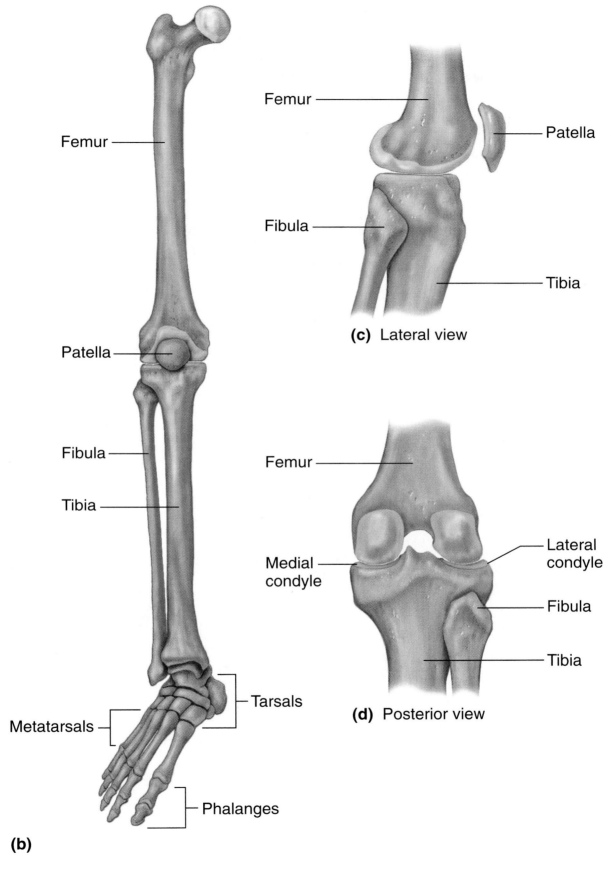

(b)

(c) Lateral view

(d) Posterior view

Figure 7.52b–d **Parts of the lower limb**

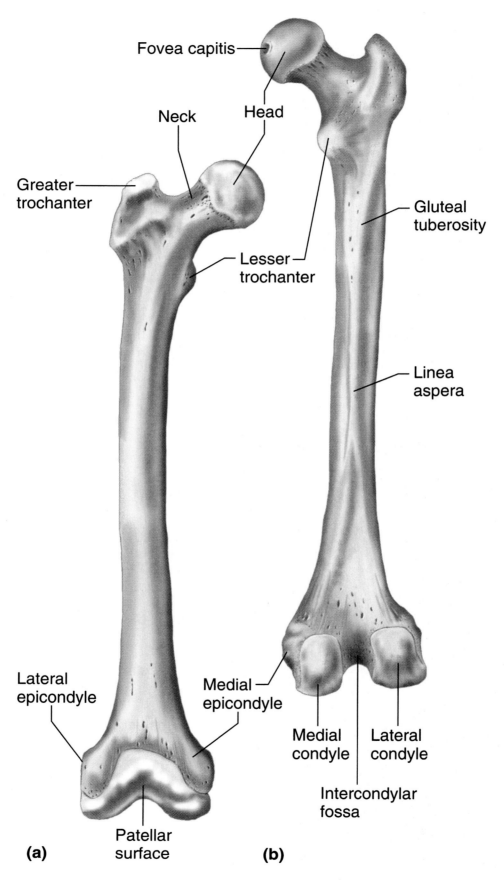

Fovea capitis

Neck   Head

Greater
trochanter

Gluteal
tuberosity

Lesser
trochanter

Linea
aspera

Lateral
epicondyle

Medial
epicondyle

Medial
condyle

Lateral
condyle

Intercondylar
fossa

(a)

Patellar
surface

(b)

Figure 7.53   Femur

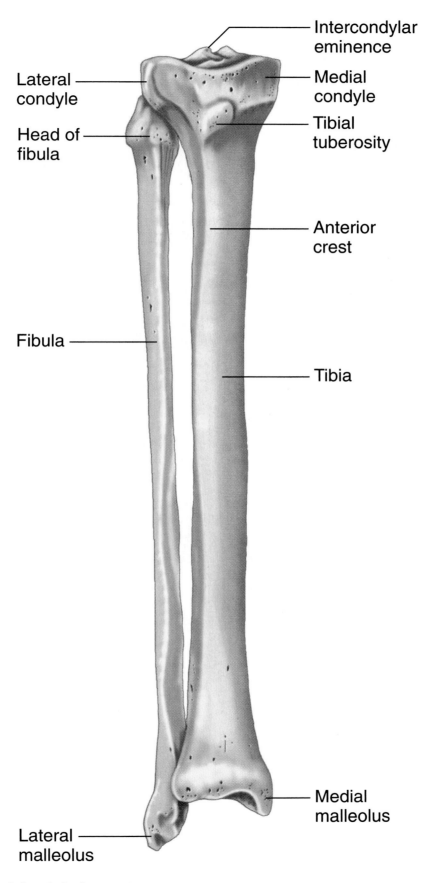

Intercondylar
eminence

Lateral
condyle

Medial
condyle

Head of
fibula

Tibial
tuberosity

Anterior
crest

Fibula

Tibia

Medial
malleolus

Lateral
malleolus

Figure 7.54   Bones of the right leg

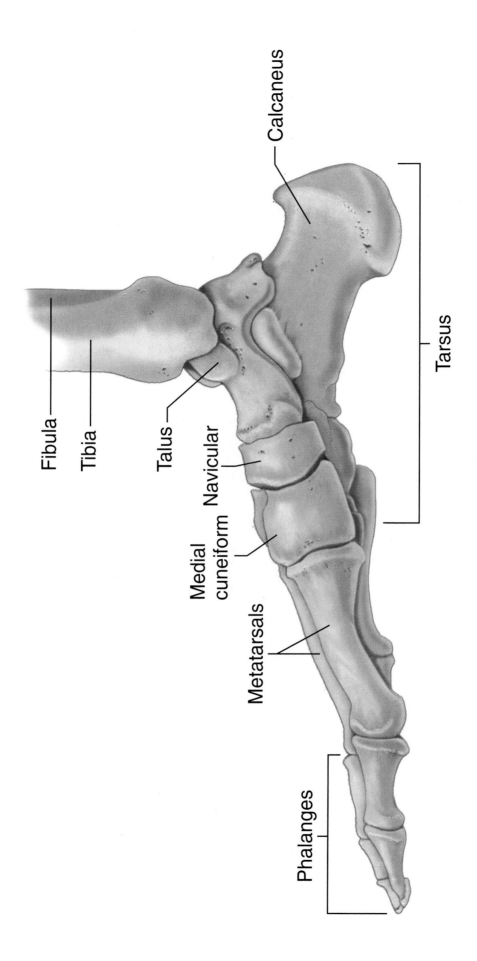

Figure 7.55b   The talus moves freely where it articulates with the tibia and fibula

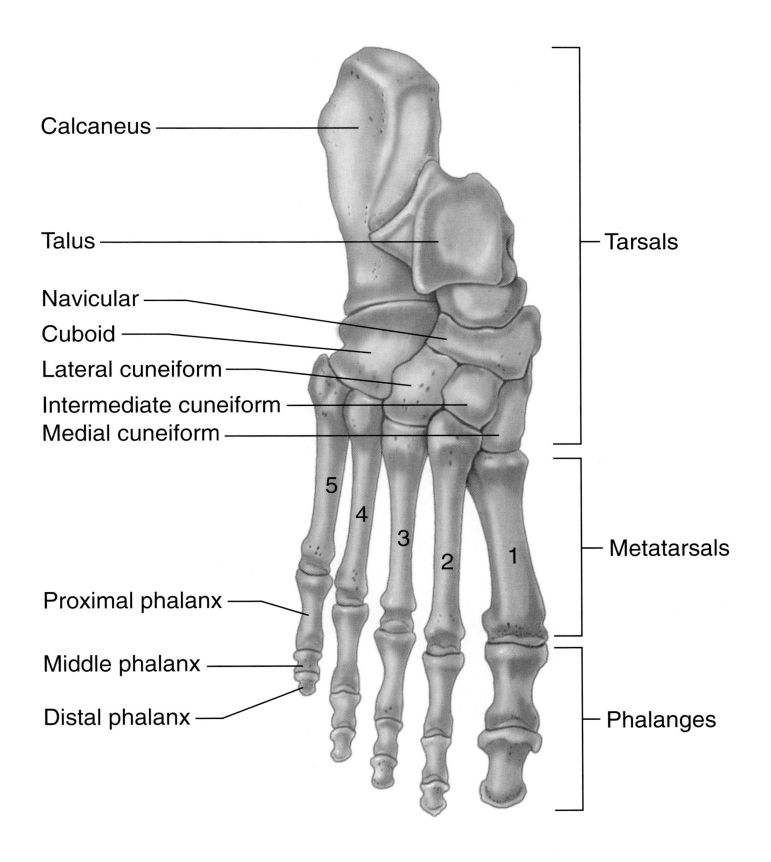

Calcaneus

Talus

Navicular

Cuboid

Lateral cuneiform

Intermediate cuneiform

Medial cuneiform

Tarsals

5

4

3

2

1

Metatarsals

Proximal phalanx

Middle phalanx

Distal phalanx

Phalanges

Figure 7.56a   **Right foot viewed superiorly**

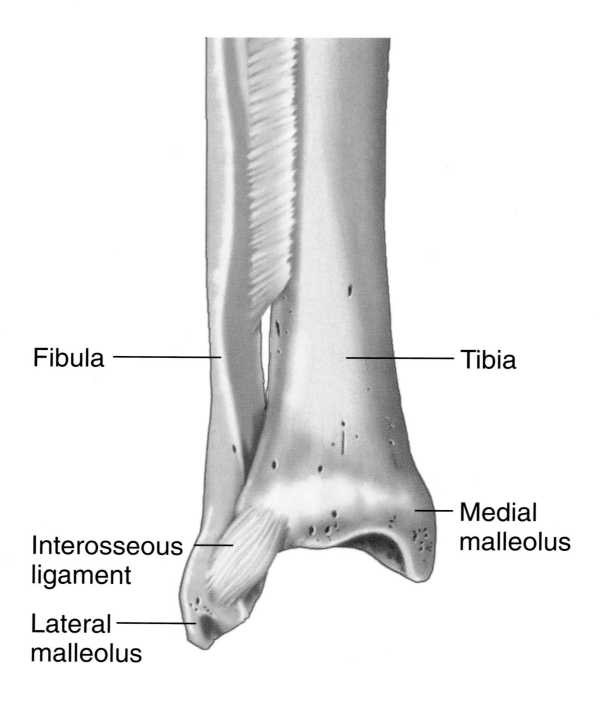

Fibula

Tibia

Interosseous
ligament

Medial
malleolus

Lateral
malleolus

Figure 8.1   An example of syndesmosis

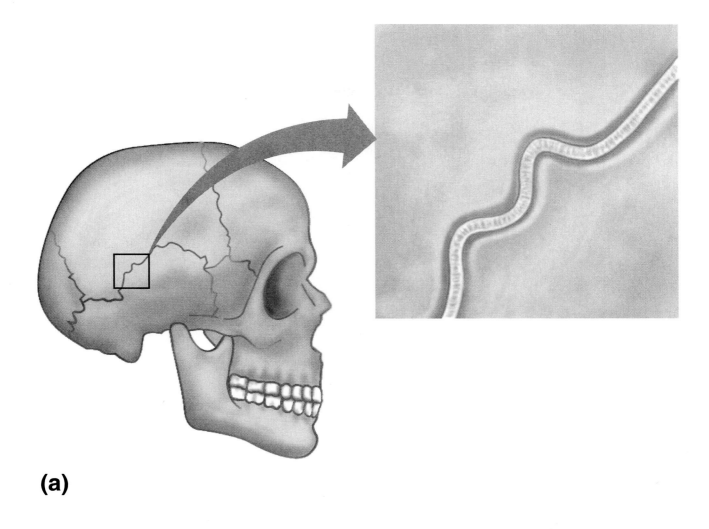

(a)

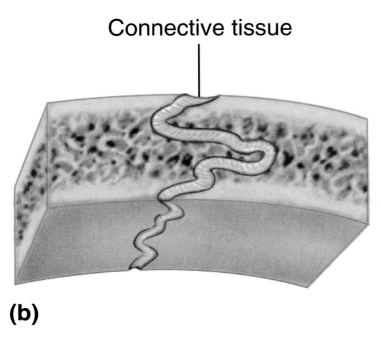

Connective tissue

(b)

Figure 8.2   Fibrous joints

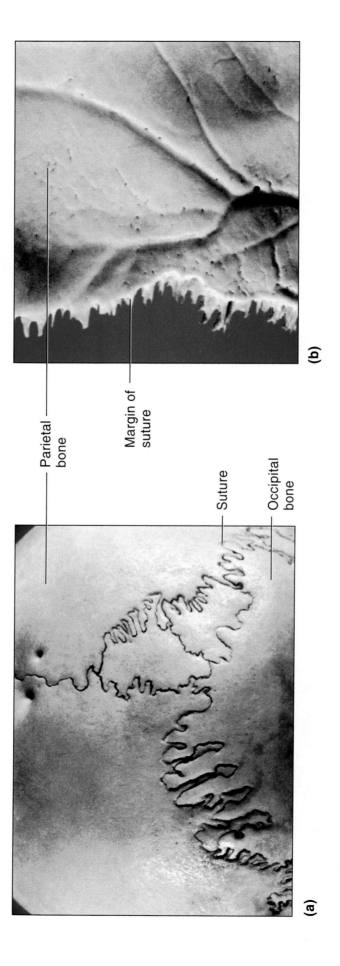

Parietal
bone

Margin of
suture

Suture

Occipital
bone

(b)

(a)

Figure 8.3 Cranial sutures

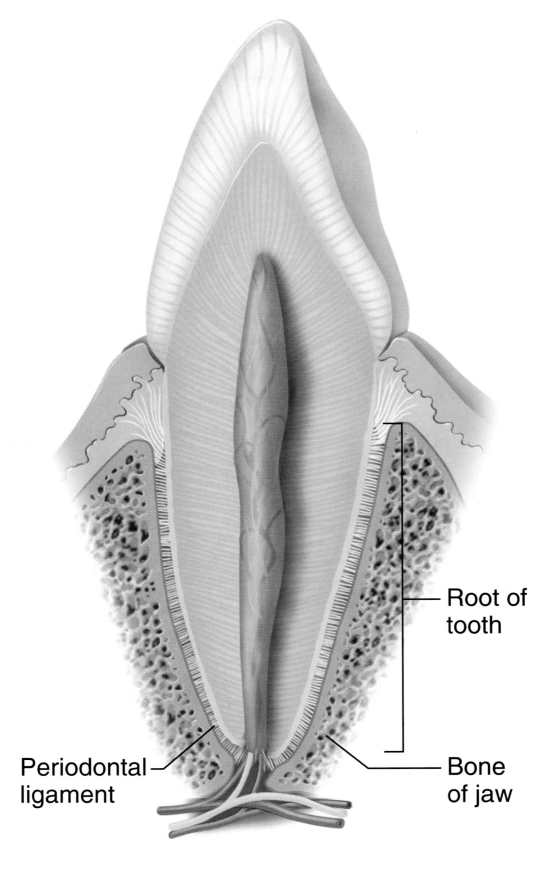

Root of
tooth

Periodontal
ligament

Bone
of jaw

Figure 8.4   The articulation between the root of a tooth and the jawbone is a gomphosis

First rib

Costal cartilage

Manubrium

Thoracic vertebra

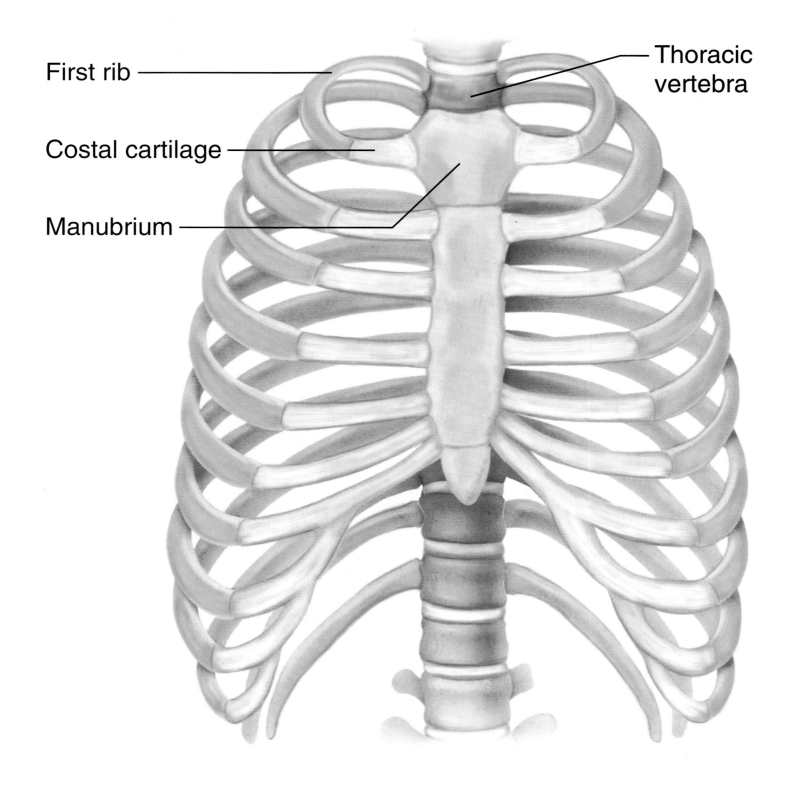

Figure 8.5   The articulation between the first rib and the manubrium is a synchondrosis

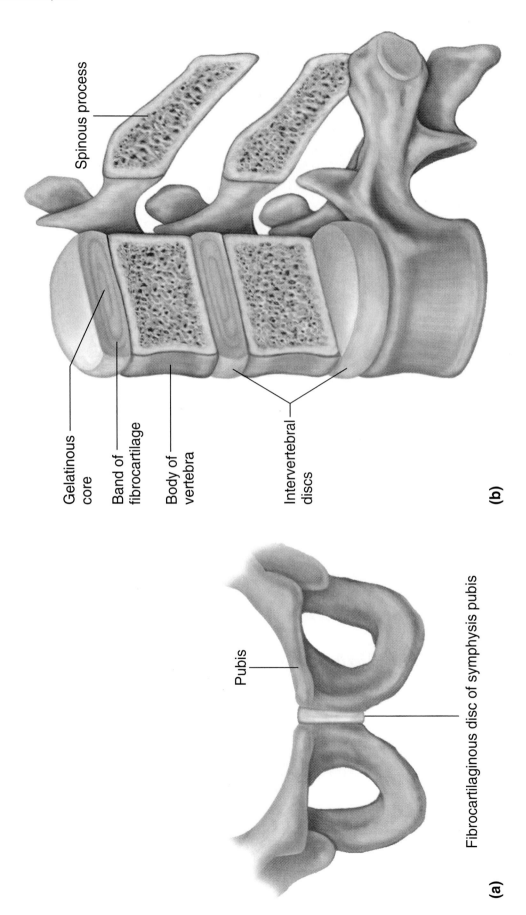

Spinous process

Gelatinous core

Band of fibrocartilage

Body of vertebra

Intervertebral discs

(b)

Pubis

Fibrocartilaginous disc of symphysis pubis

(a)

Figure 8.6   Fibrocartilage composes the symphysis pubis and the intervertebral discs

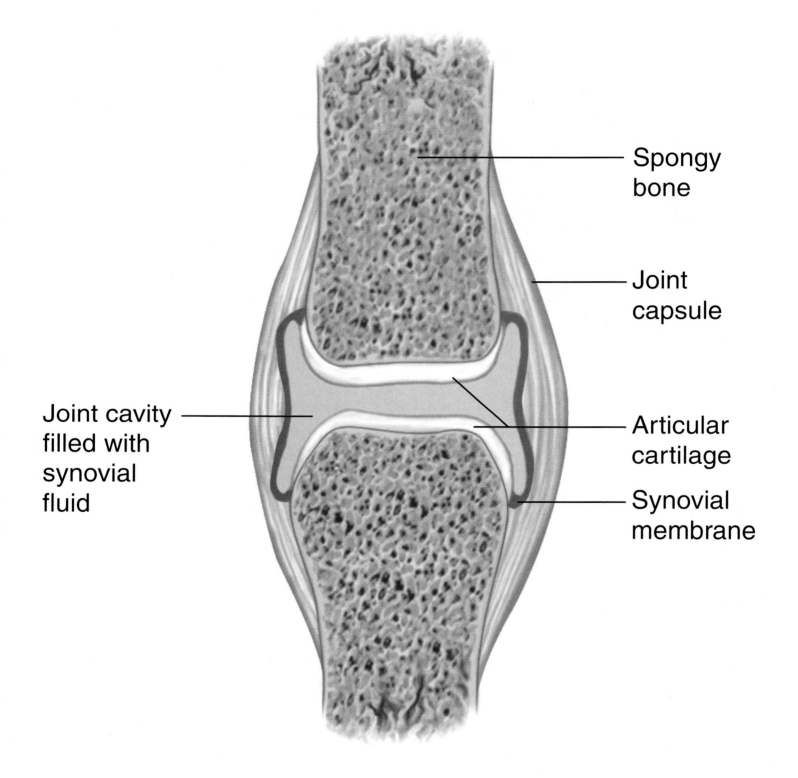

Spongy bone

Joint capsule

Joint cavity filled with synovial fluid

Articular cartilage

Synovial membrane

Figure 8.7   The generalized structure of a synovial joint

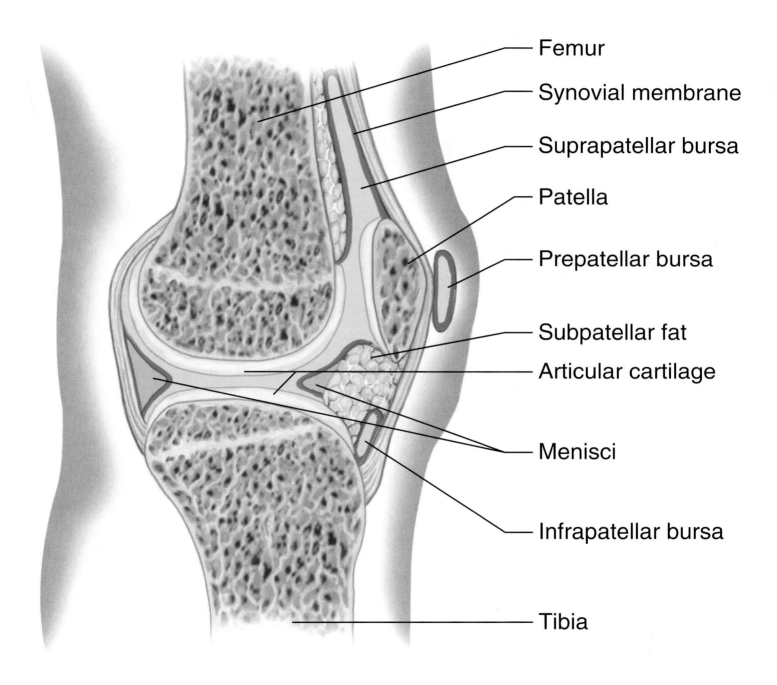

Figure 8.8   **Menisci separate the articulating surfaces of the femur and tibia**

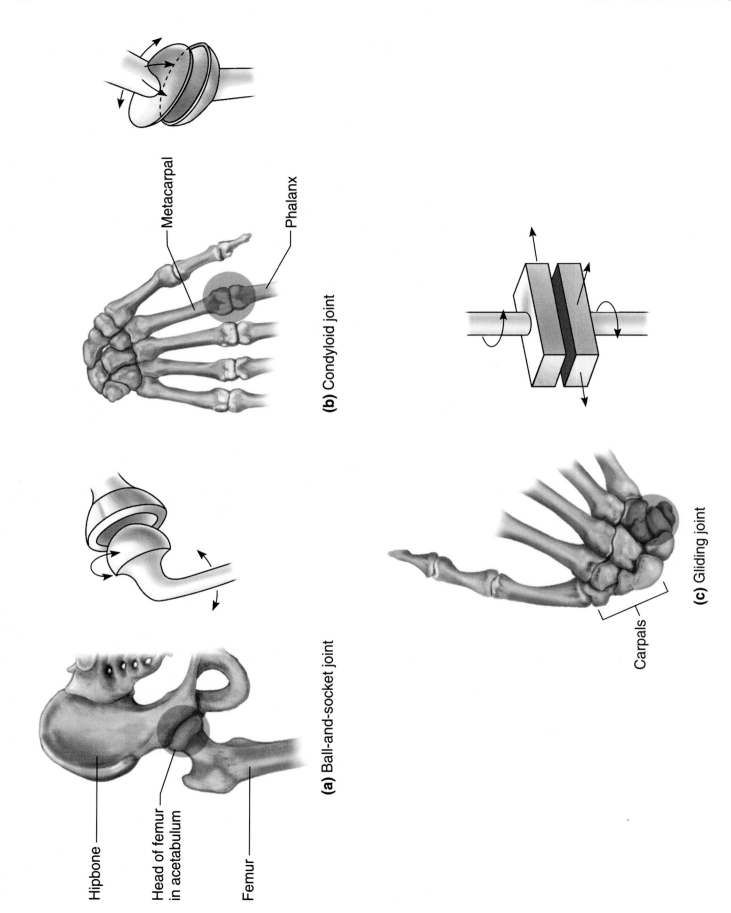

(a) Ball-and-socket joint

Hipbone

Head of femur in acetabulum

Femur

(b) Condyloid joint

Metacarpal

Phalanx

(c) Gliding joint

Carpals

Figure 8.9a–c Types and examples of synovial (freely movable) joints

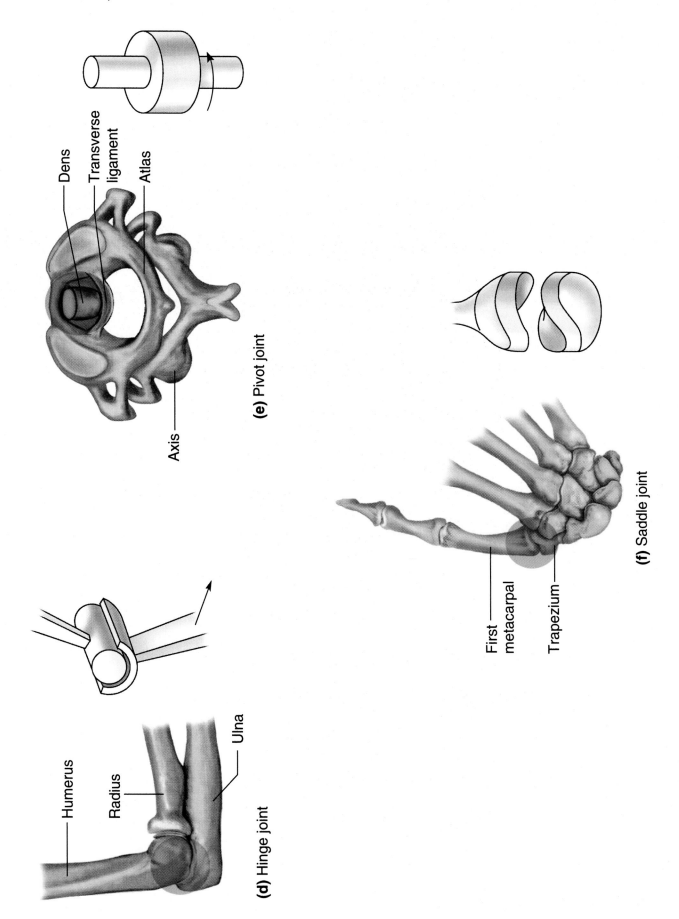

Dens

Transverse ligament

Atlas

Axis

(e) Pivot joint

(f) Saddle joint

First metacarpal

Trapezium

Humerus

Radius

Ulna

(d) Hinge joint

Figure 8.9d–f Types and examples of synovial (freely movable) joints

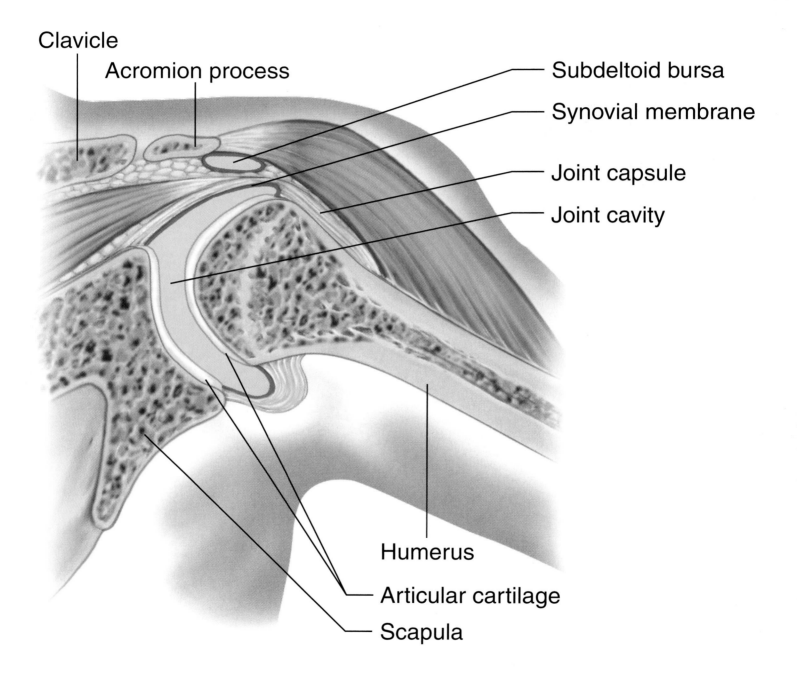

Figure 8.13a   **The shoulder joint allows movement in all directions**

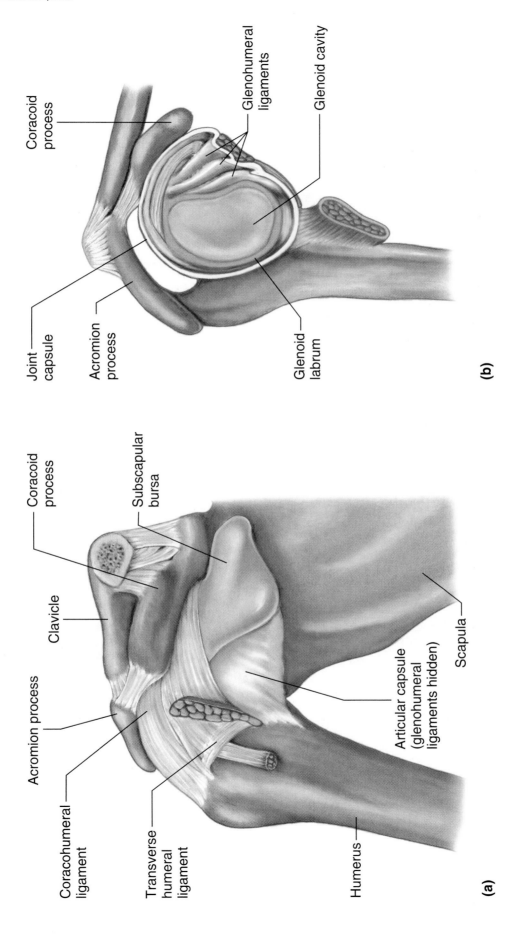

Figure 8.14 Ligaments associated with the shoulder joint

(a)

Acromion process

Coracoid process

Clavicle

Subscapular bursa

Coracohumeral ligament

Transverse humeral ligament

Articular capsule (glenohumeral ligaments hidden)

Scapula

Humerus

(b)

Coracoid process

Glenohumeral ligaments

Glenoid cavity

Joint capsule

Acromion process

Glenoid labrum

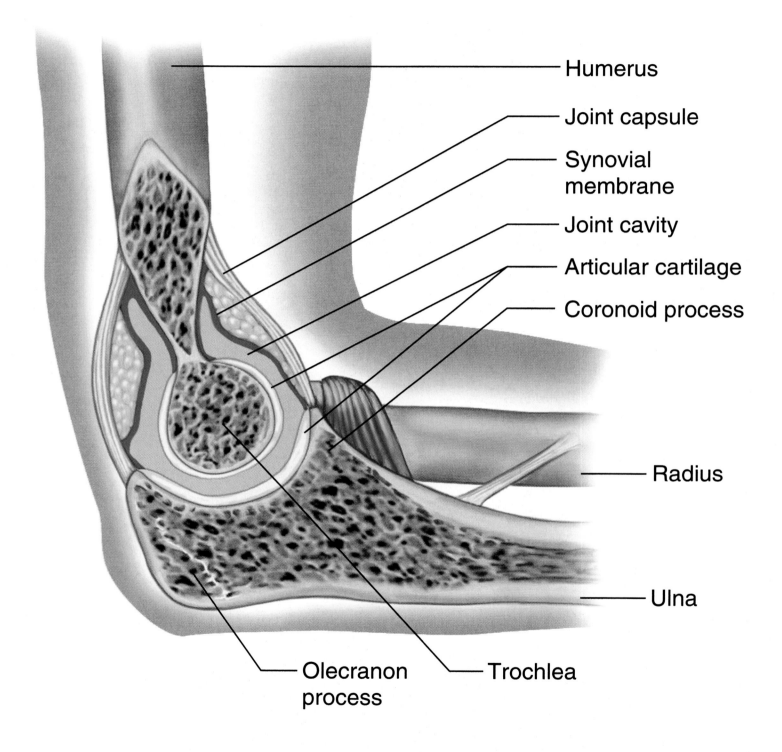

Humerus

Joint capsule

Synovial membrane

Joint cavity

Articular cartilage

Coronoid process

Radius

Ulna

Olecranon process

Trochlea

Figure 8.15a   The elbow joint allows hinge movements and pronation and supination of the hand

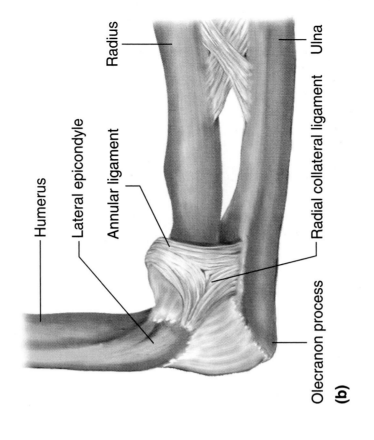

Radius

Ulna

Humerus

Lateral epicondyle

Annular ligament

Radial collateral ligament

Olecranon process

(b)

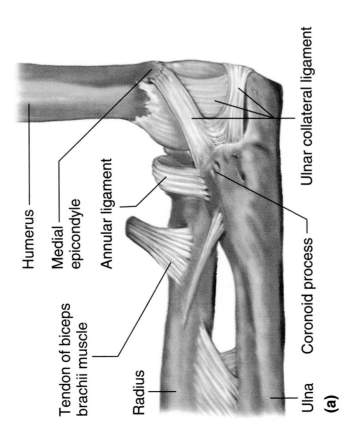

Humerus

Medial epicondyle

Annular ligament

Tendon of biceps brachii muscle

Radius

Ulna

Ulnar collateral ligament

Coronoid process

(a)

Figure 8.16   Ligaments associated with the elbow joint

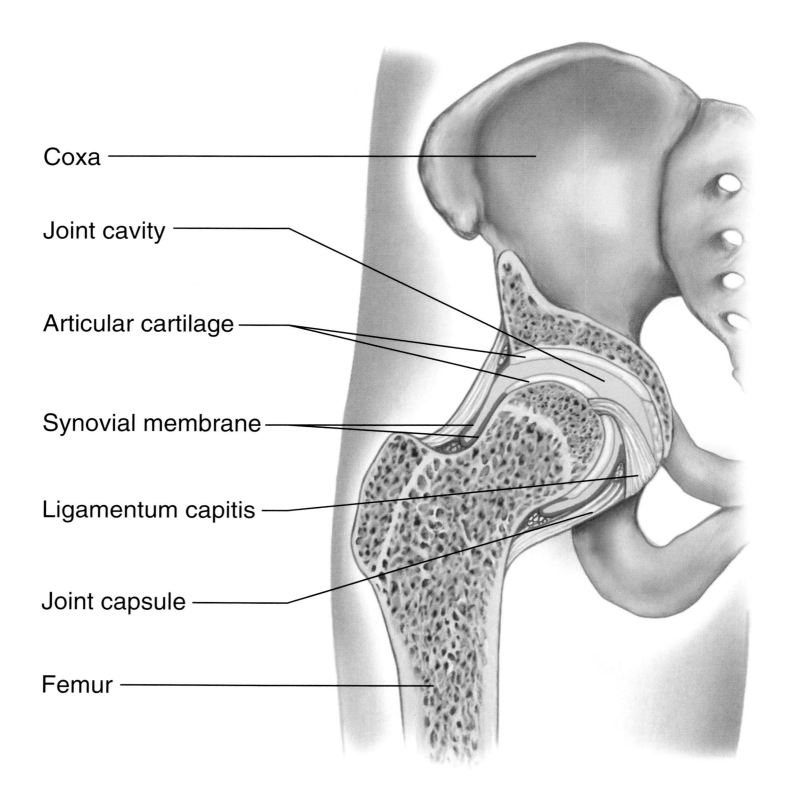

Coxa

Joint cavity

Articular cartilage

Synovial membrane

Ligamentum capitis

Joint capsule

Femur

Figure 8.18a   A ring of cartilage and a ligament-reinforced joint capsule hold together the hip joint

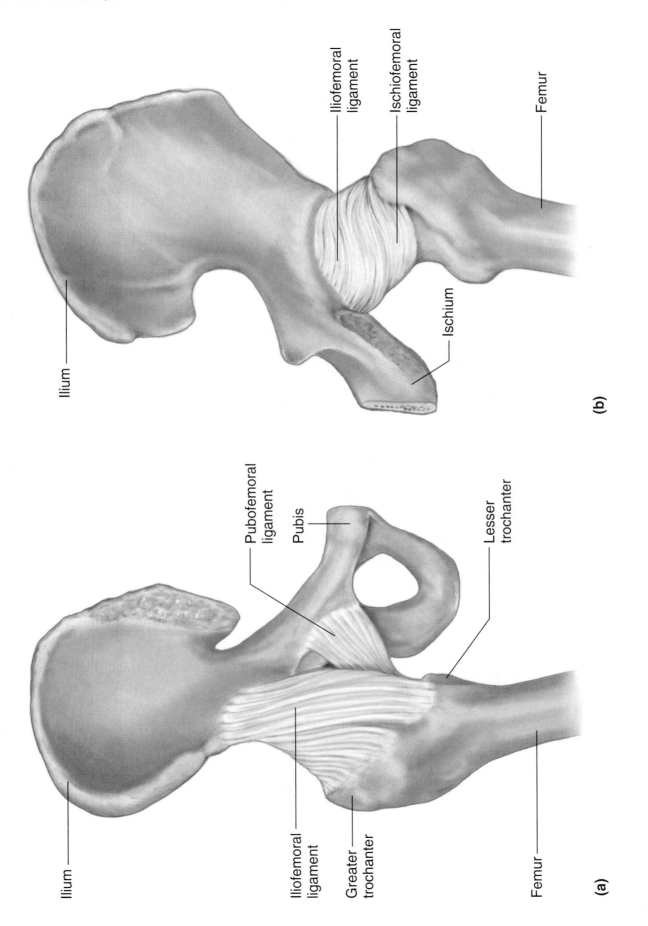

Iliofemoral ligament

Ischiofemoral ligament

Femur

Ilium

Ischium

(b)

Pubofemoral ligament

Pubis

Lesser trochanter

Ilium

Iliofemoral ligament

Greater trochanter

Femur

(a)

Figure 8.19 The major ligaments of the right hip joint

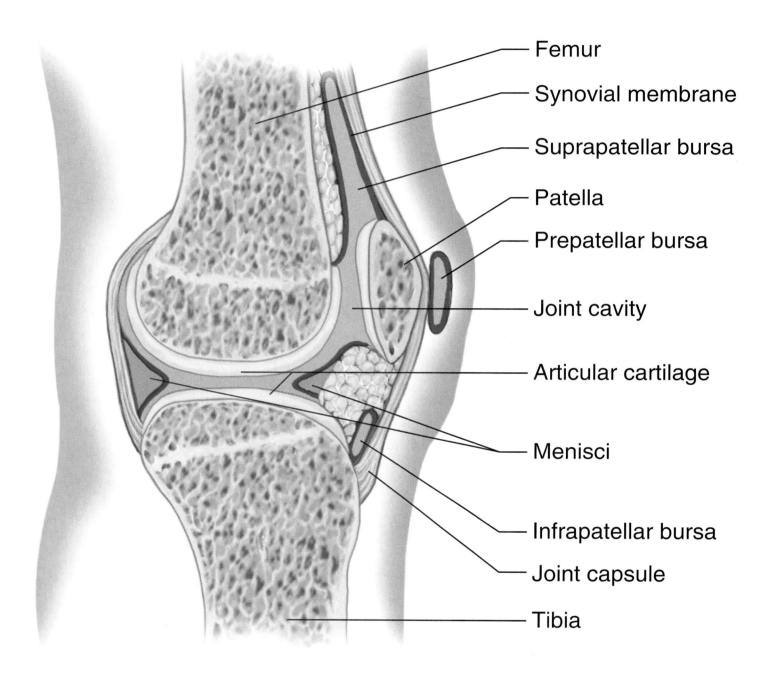

Femur

Synovial membrane

Suprapatellar bursa

Patella

Prepatellar bursa

Joint cavity

Articular cartilage

Menisci

Infrapatellar bursa

Joint capsule

Tibia

Figure 8.20a   The knee joint is the most complex of the synovial joints

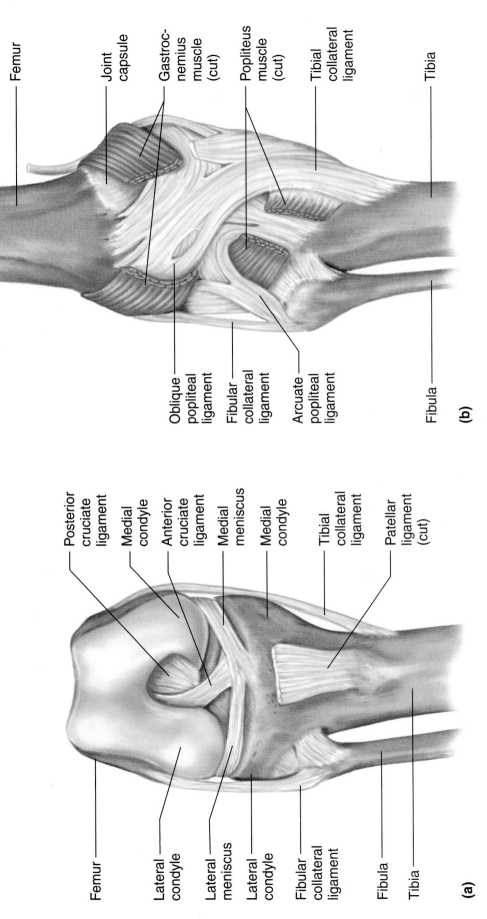

Figure 8.21 Ligaments within the knee joint help to strengthen it

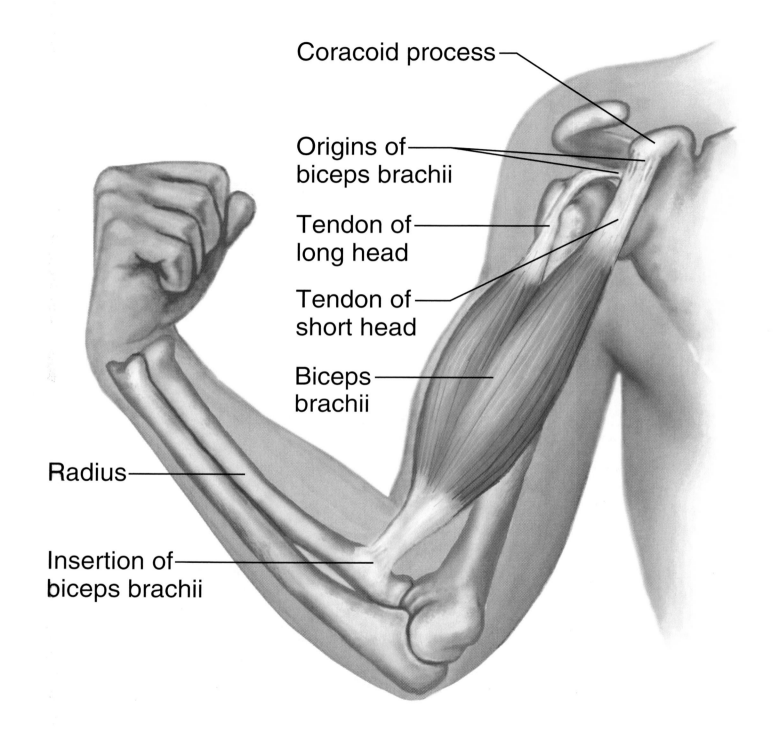

Coracoid process

Origins of biceps brachii

Tendon of long head

Tendon of short head

Biceps brachii

Radius

Insertion of biceps brachii

Figure 9.19   The biceps brachii has two heads that originate on the scapula

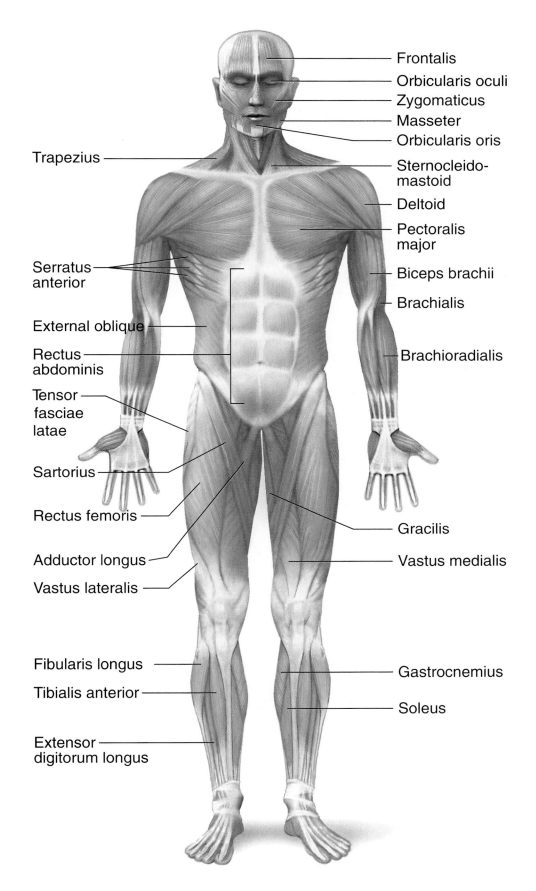

Frontalis

Orbicularis oculi

Zygomaticus

Masseter

Orbicularis oris

Trapezius

Sternocleido-
mastoid

Deltoid

Pectoralis
major

Serratus
anterior

Biceps brachii

Brachialis

External oblique

Rectus
abdominis

Brachioradialis

Tensor
fasciae
latae

Sartorius

Rectus femoris

Gracilis

Adductor longus

Vastus medialis

Vastus lateralis

Fibularis longus

Gastrocnemius

Tibialis anterior

Soleus

Extensor
digitorum longus

Figure 9.20   Anterior view of superficial skeletal muscles

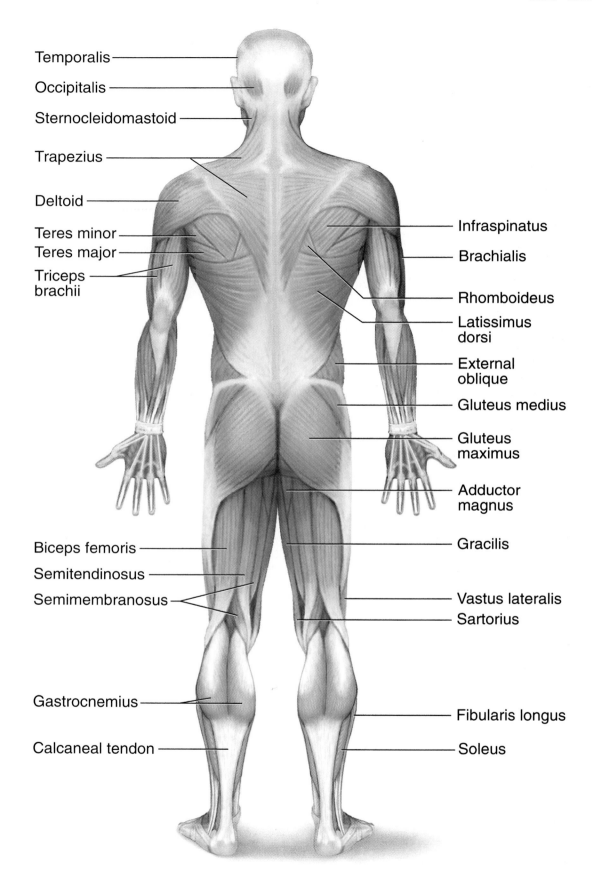

Temporalis

Occipitalis

Sternocleidomastoid

Trapezius

Deltoid

Teres minor

Teres major

Triceps brachii

Infraspinatus

Brachialis

Rhomboideus

Latissimus dorsi

External oblique

Gluteus medius

Gluteus maximus

Adductor magnus

Gracilis

Biceps femoris

Semitendinosus

Semimembranosus

Vastus lateralis

Sartorius

Gastrocnemius

Calcaneal tendon

Fibularis longus

Soleus

Figure 9.21   Posterior view of superficial skeletal muscles

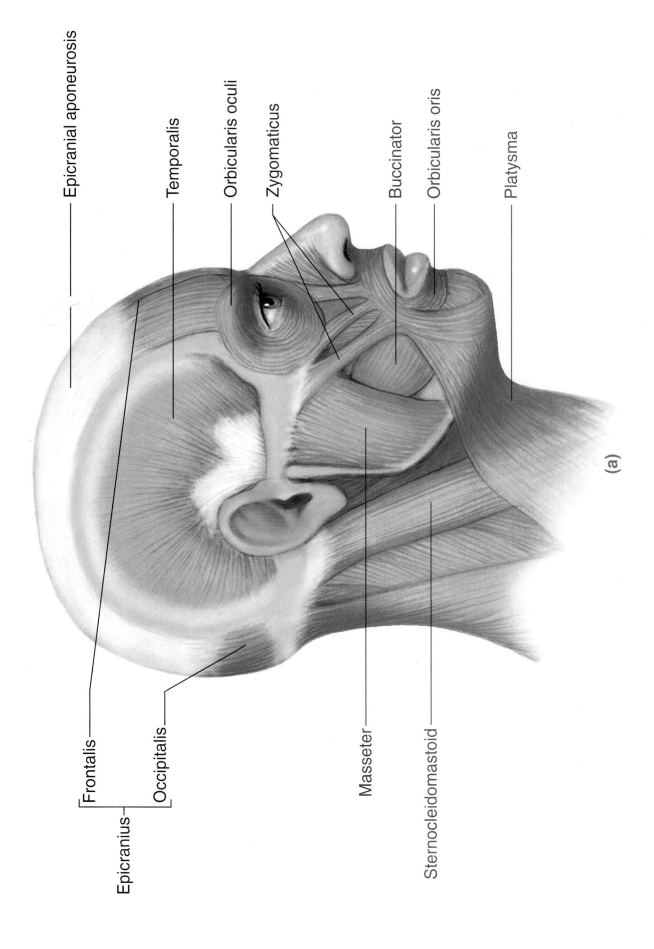

Epicranial aponeurosis

Temporalis

Orbicularis oculi

Zygomaticus

Buccinator

Orbicularis oris

Platysma

Frontalis

Occipitalis

Epicranius

Masseter

Sternocleidomastoid

(a)

Figure 9.22a Muscles of facial expression and mastication

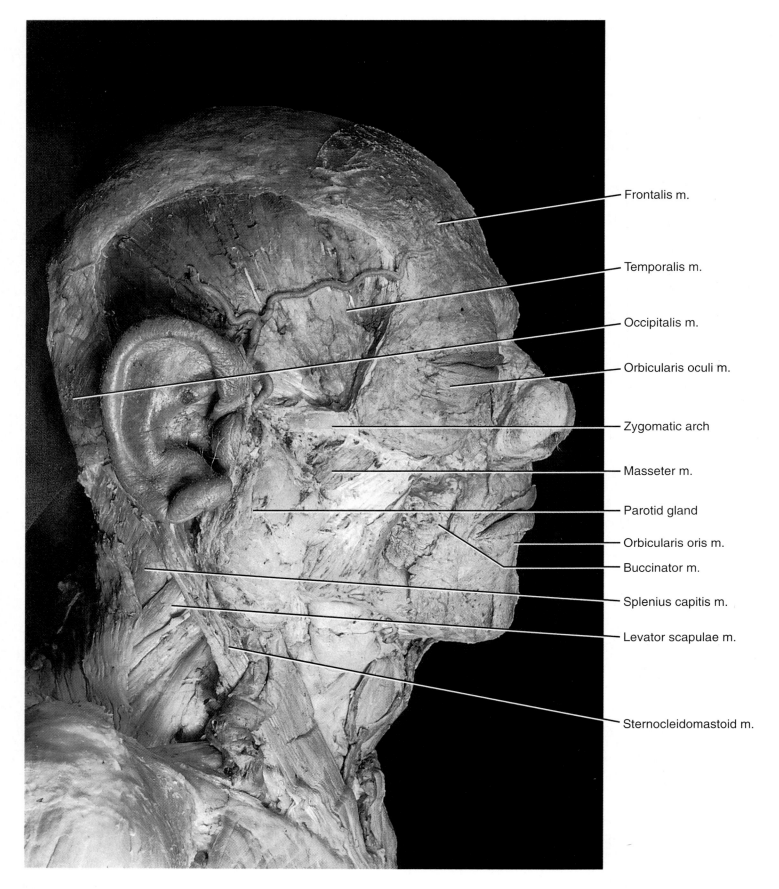

Frontalis m.

Temporalis m.

Occipitalis m.

Orbicularis oculi m.

Zygomatic arch

Masseter m.

Parotid gland

Orbicularis oris m.

Buccinator m.

Splenius capitis m.

Levator scapulae m.

Sternocleidomastoid m.

Plate 61   Lateral view of the head

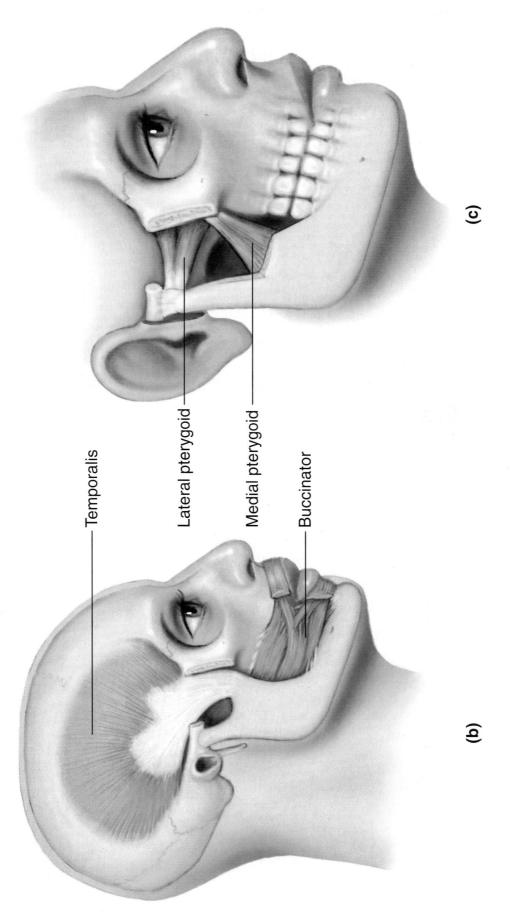

Temporalis

Lateral pterygoid

Medial pterygoid

Buccinator

(c)

(b)

Figure 9.22b–c Temporalis and buccinator muscles and the lateral and medial pterygoid muscles

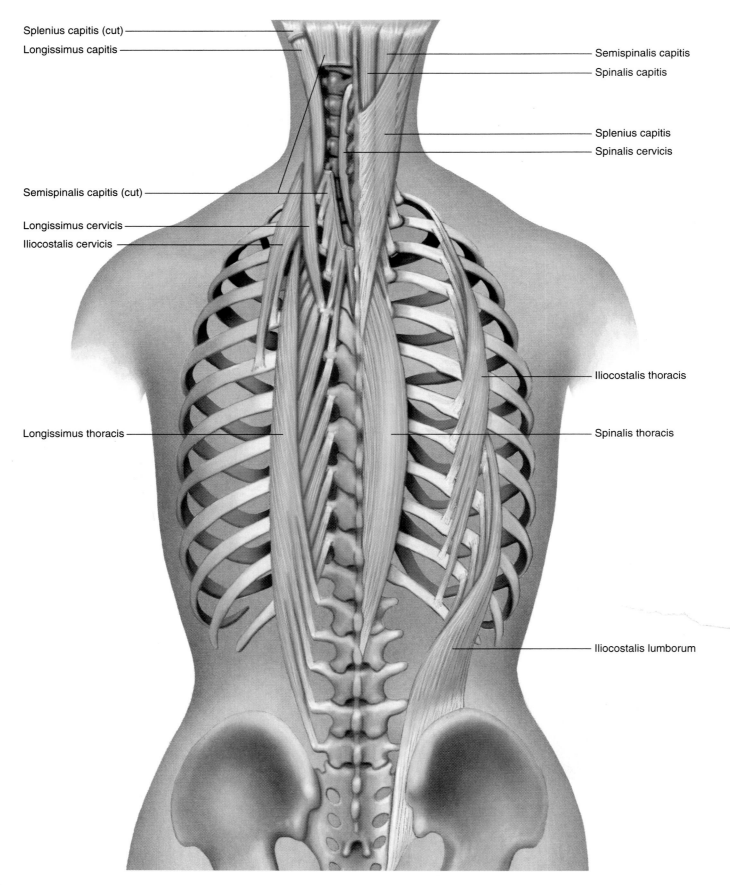

Splenius capitis (cut)

Longissimus capitis

Semispinalis capitis

Spinalis capitis

Splenius capitis

Spinalis cervicis

Semispinalis capitis (cut)

Longissimus cervicis

Iliocostalis cervicis

Iliocostalis thoracis

Longissimus thoracis

Spinalis thoracis

Iliocostalis lumborum

Figure 9.23   Deep muscles of the back and the neck help move the head and hold the torso erect

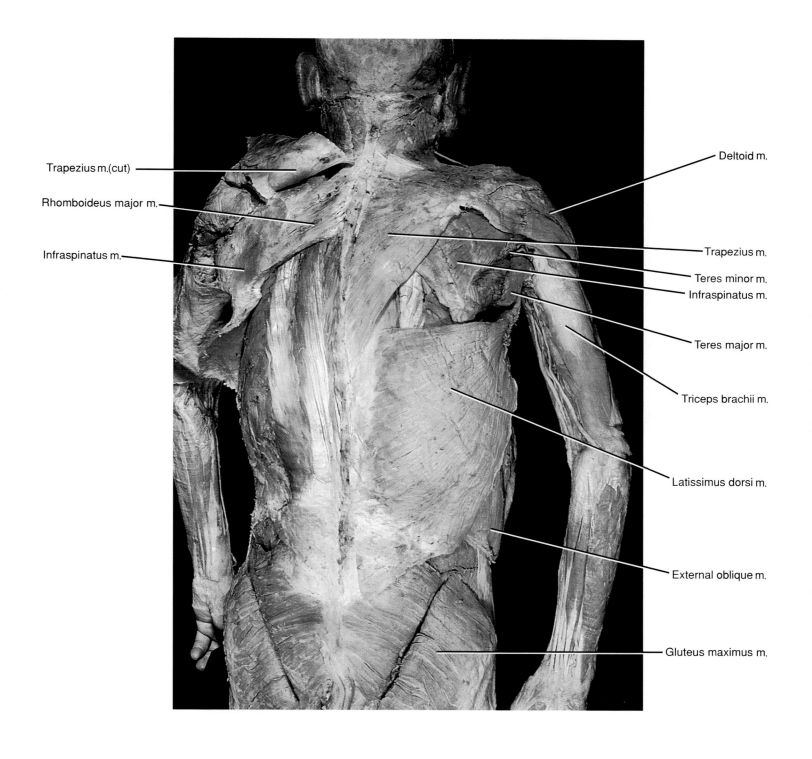

Trapezius m.(cut)

Rhomboideus major m.

Infraspinatus m.

Deltoid m.

Trapezius m.

Teres minor m.

Infraspinatus m.

Teres major m.

Triceps brachii m.

Latissimus dorsi m.

External oblique m.

Gluteus maximus m.

Plate 63   Posterior view of the trunk, with deep thoracic muscles exposed on the left

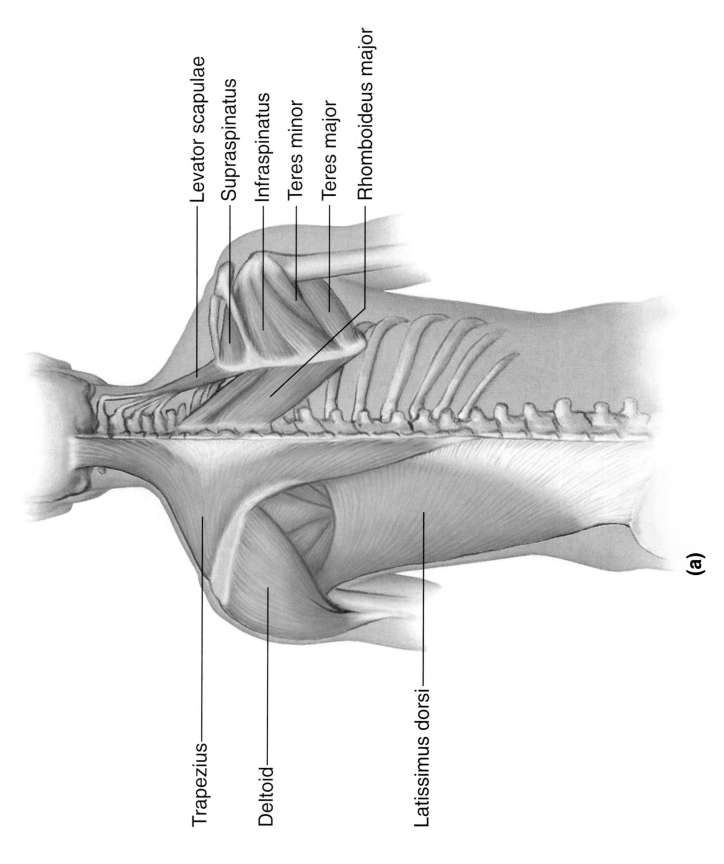

Levator scapulae

Supraspinatus

Infraspinatus

Teres minor

Teres major

Rhomboideus major

Trapezius

Deltoid

Latissimus dorsi

(a)

Figure 9.24a   Muscles of the posterior shoulder

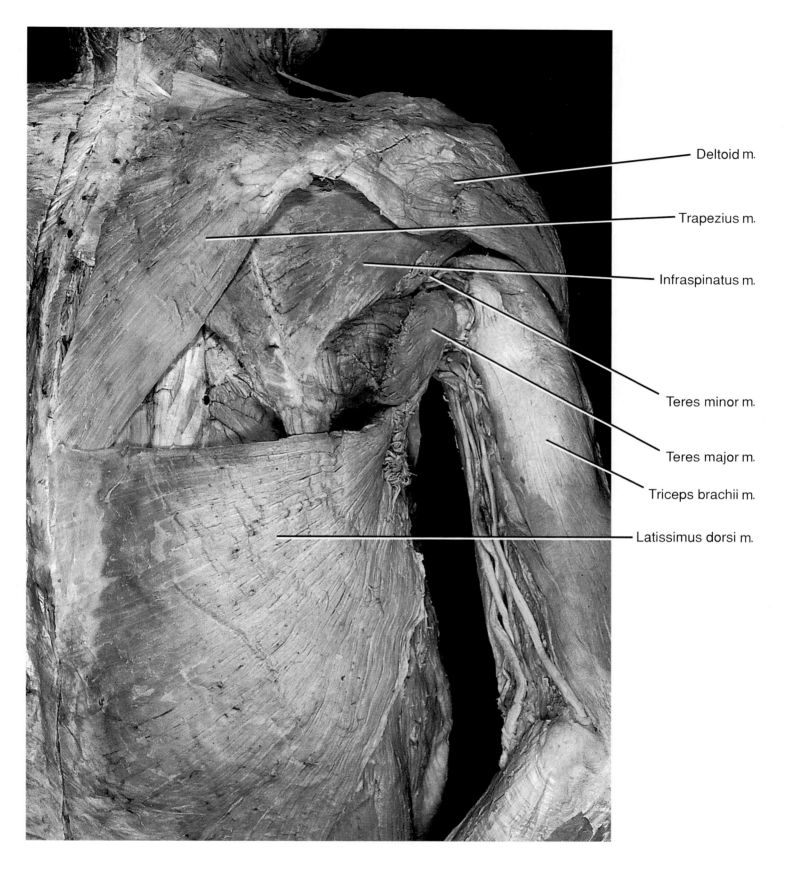

Deltoid m.

Trapezius m.

Infraspinatus m.

Teres minor m.

Teres major m.

Triceps brachii m.

Latissimus dorsi m.

Plate 64 Posterior view of the right thorax and arm

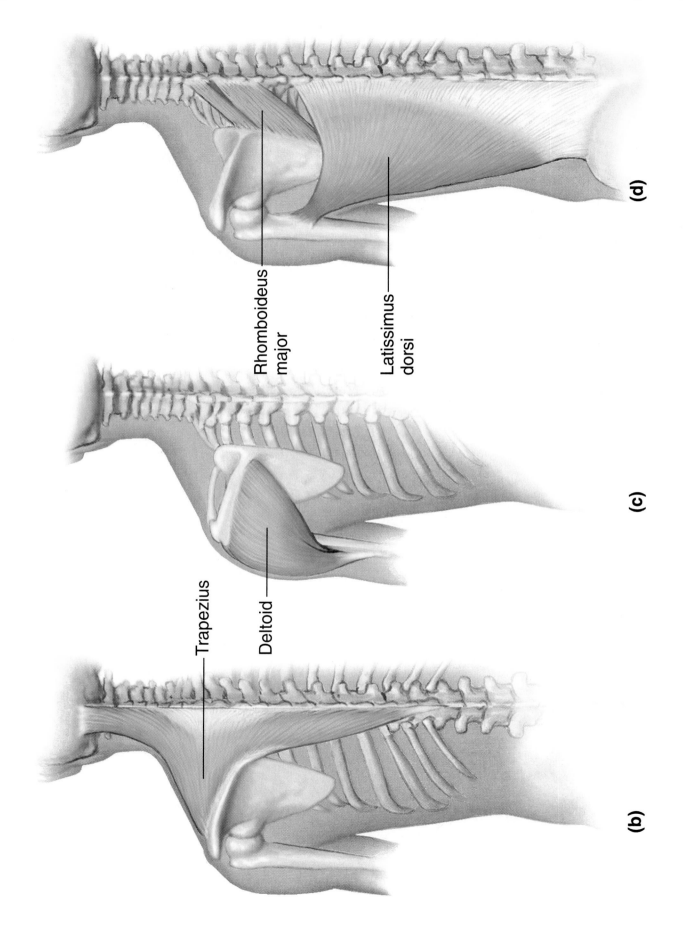

Rhomboideus major

Latissimus dorsi

(d)

Trapezius

Deltoid

(c)

(b)

Figure 9.24b–d Views of the trapezius, deltoid, and rhomboideus and latissimus dorsi muscles

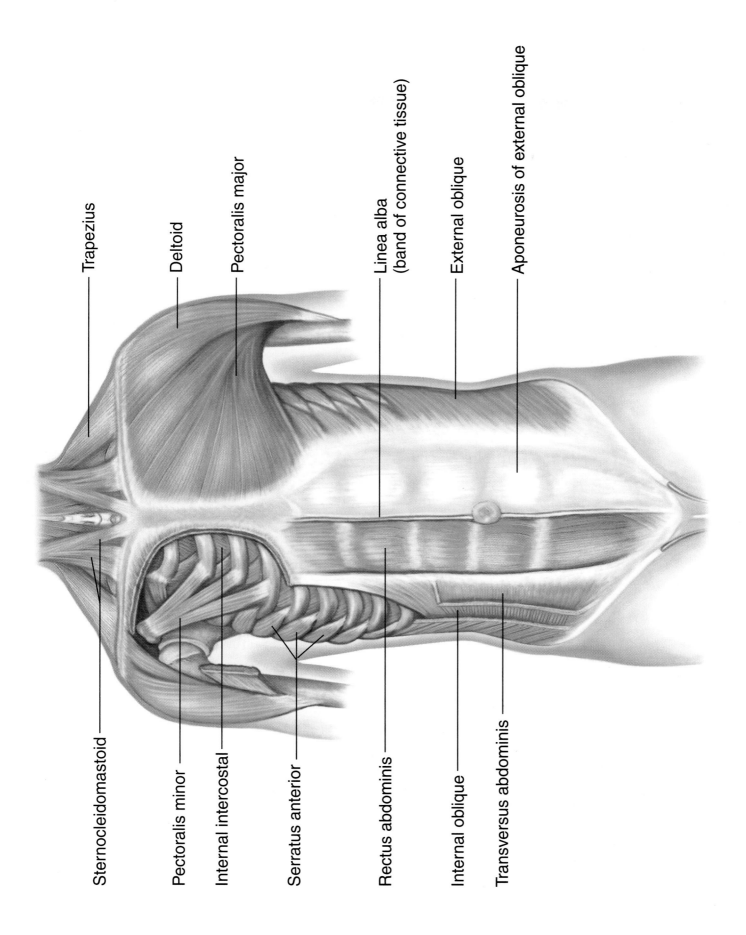

Trapezius

Deltoid

Pectoralis major

Linea alba
(band of connective tissue)

External oblique

Aponeurosis of external oblique

Sternocleidomastoid

Pectoralis minor

Internal intercostal

Serratus anterior

Rectus abdominis

Internal oblique

Transversus abdominis

Figure 9.25 Muscles of the anterior chest and abdominal wall

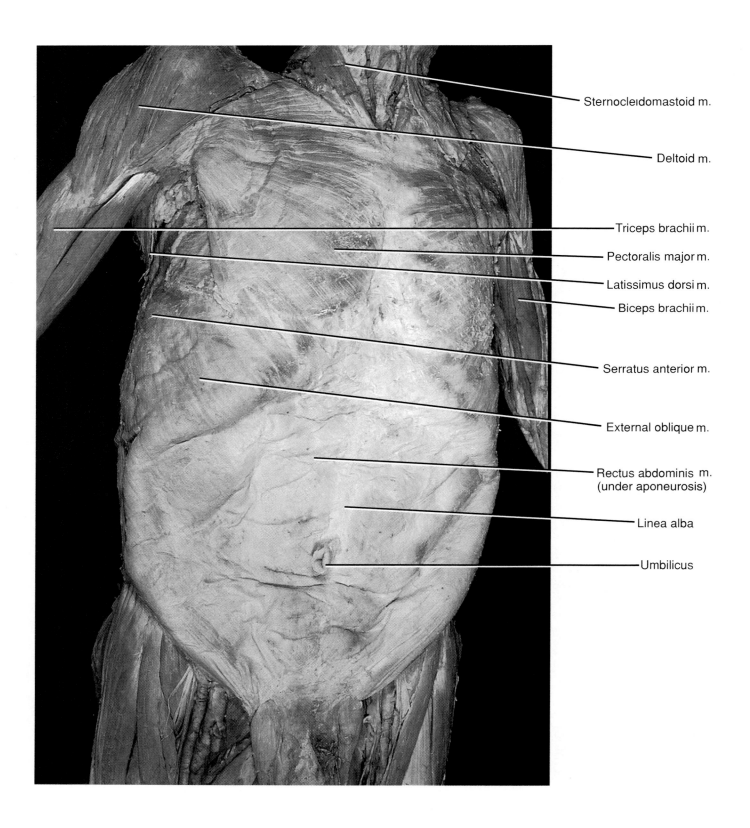

Sternocleidomastoid m.

Deltoid m.

Triceps brachii m.

Pectoralis major m.

Latissimus dorsi m.

Biceps brachii m.

Serratus anterior m.

External oblique m.

Rectus abdominis m.
(under aponeurosis)

Linea alba

Umbilicus

Plate 62   Anterior view of the trunk

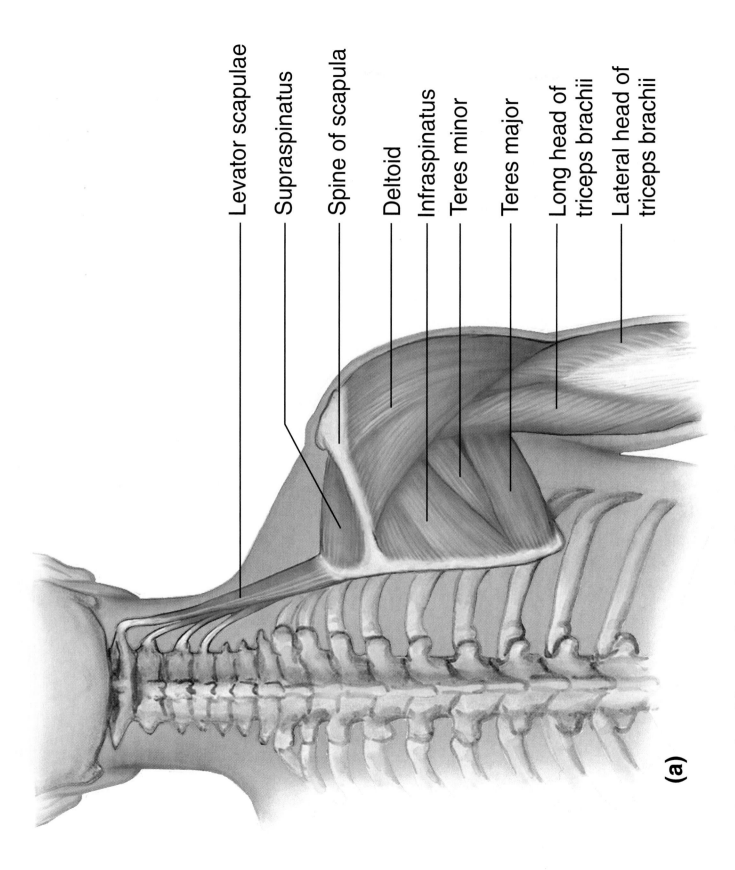

Levator scapulae

Supraspinatus

Spine of scapula

Deltoid

Infraspinatus

Teres minor

Teres major

Long head of triceps brachii

Lateral head of triceps brachii

(a)

Figure 9.26a Muscles of the posterior surface of the scapula and arm

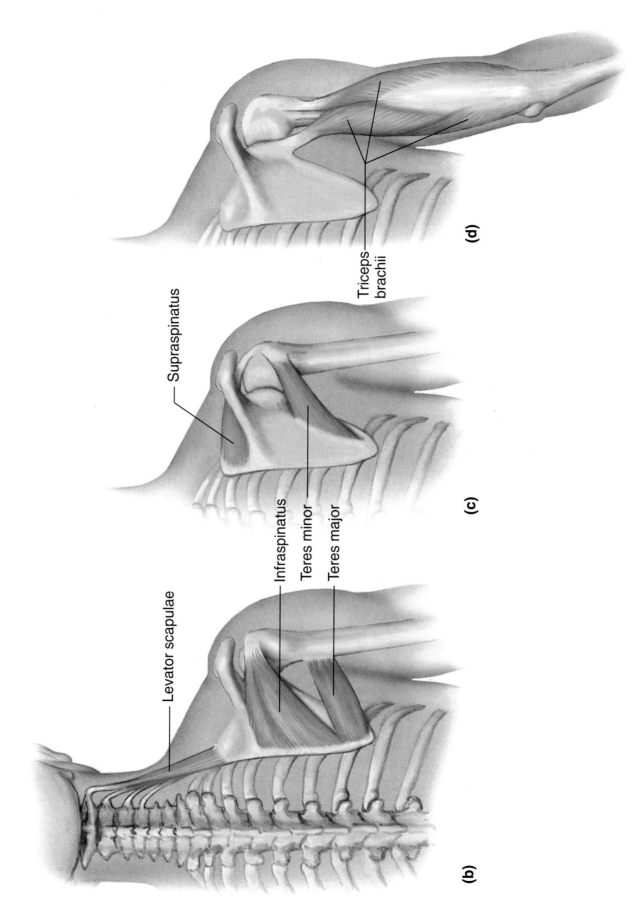

(d)

Triceps
brachii

Supraspinatus

(c)

Infraspinatus

Teres minor

Teres major

Levator scapulae

(b)

Figure 9.26b–d   Muscles associated with the scapula, and isolated view of the triceps brachii

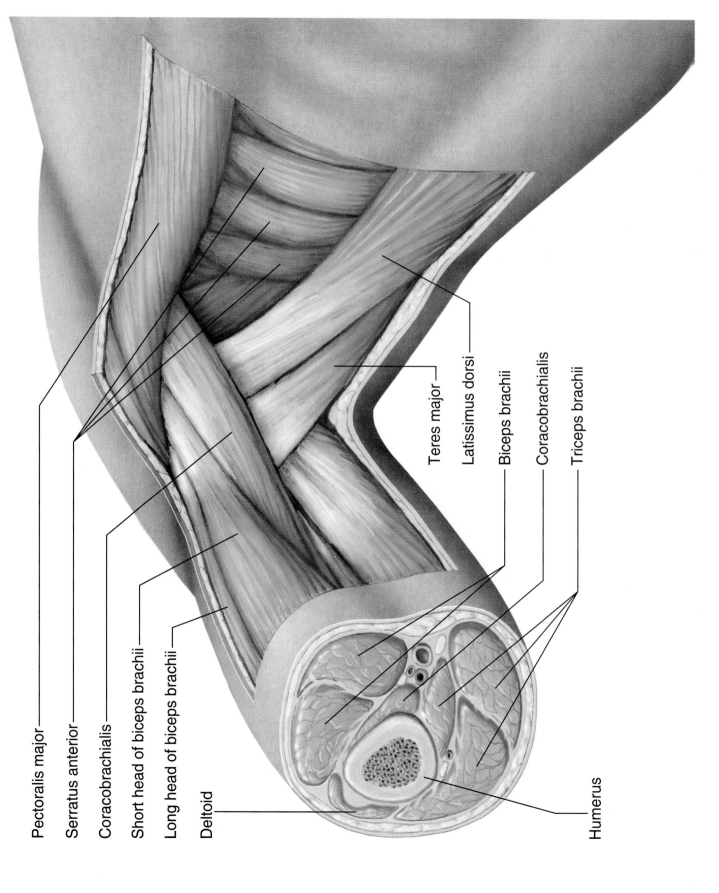

Pectoralis major

Serratus anterior

Coracobrachialis

Short head of biceps brachii

Long head of biceps brachii

Deltoid

Teres major

Latissimus dorsi

Biceps brachii

Coracobrachialis

Triceps brachii

Humerus

Figure 9.27   Cross section of the arm

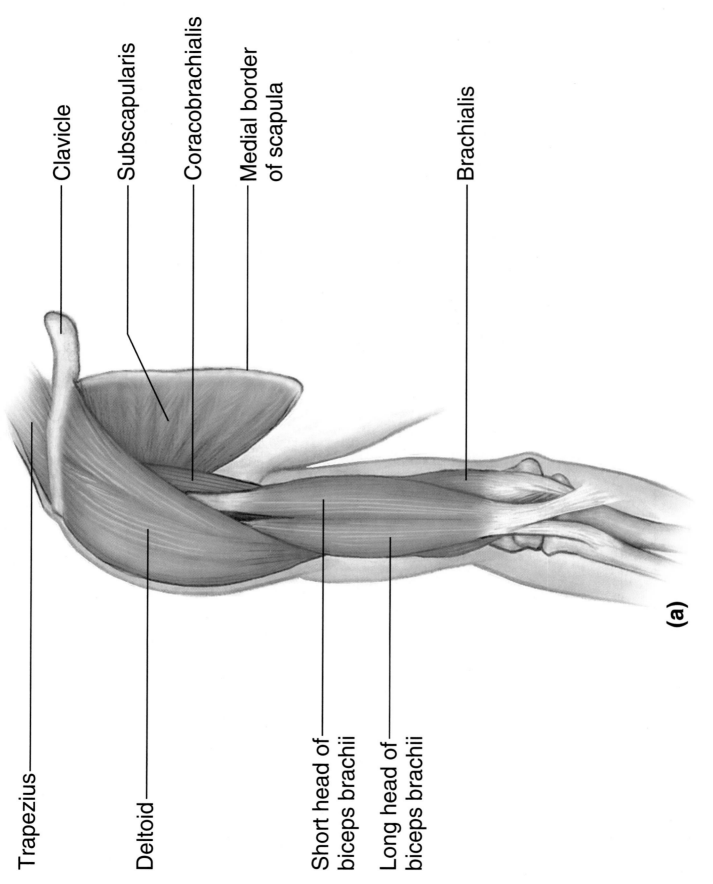

Trapezius

Clavicle

Subscapularis

Coracobrachialis

Medial border of scapula

Brachialis

Deltoid

Short head of biceps brachii

Long head of biceps brachii

(a)

Figure 9.28a Muscles of the anterior shoulder and the arm, with the rib cage removed

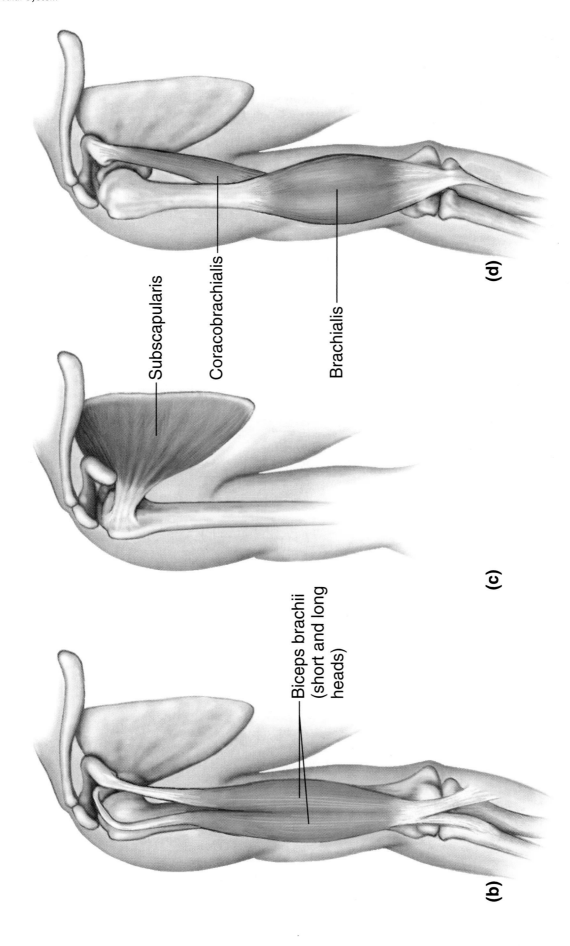

Subscapularis

Coracobrachialis

Brachialis

Biceps brachii (short and long heads)

(d)

(c)

(b)

Figure 9.28b–d Isolated views of muscles associated with the arm

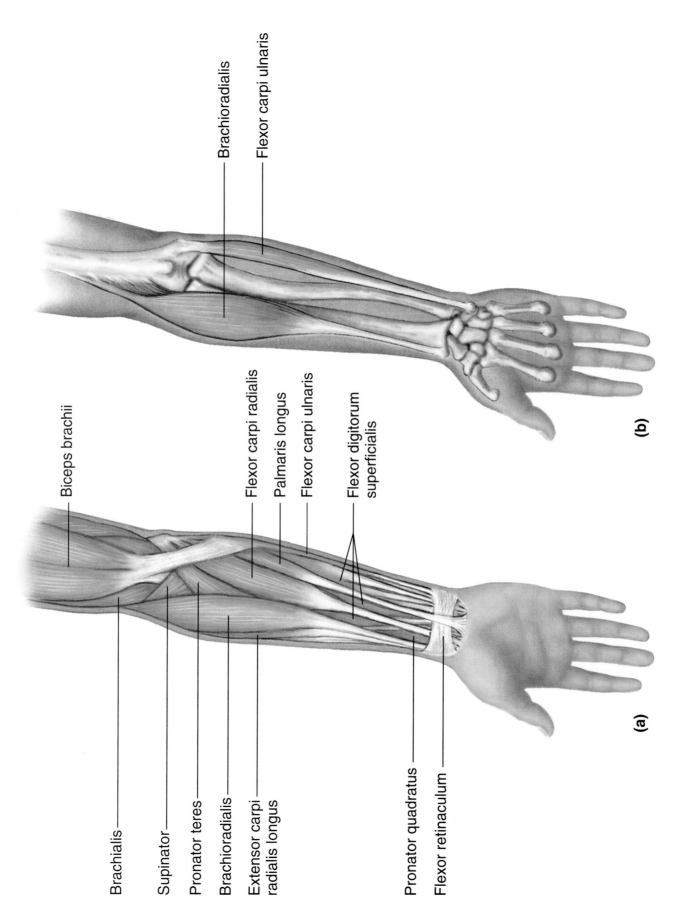

Biceps brachii

Brachioradialis

Flexor carpi ulnaris

Flexor carpi radialis

Palmaris longus

Flexor carpi ulnaris

Flexor digitorum
superficialis

Brachialis

Supinator

Pronator teres

Brachioradialis

Extensor carpi
radialis longus

Pronator quadratus

Flexor retinaculum

(a)

(b)

Figure 9.29a–b   Muscles of the anterior forearm

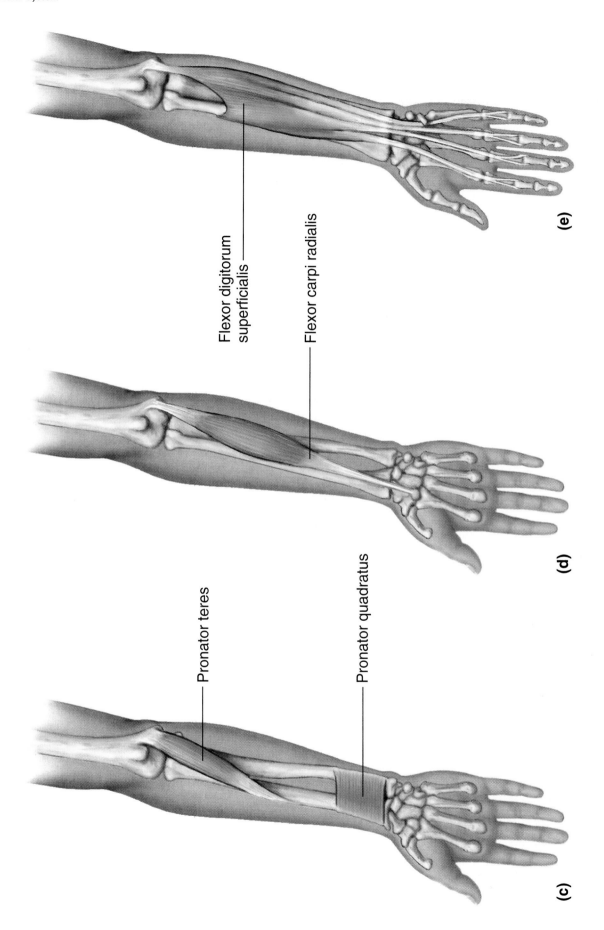

Pronator teres

Flexor digitorum superficialis

Pronator quadratus

Flexor carpi radialis

(c)

(d)

(e)

Figure 9.29c–e Isolated views of muscles associated with the anterior forearm

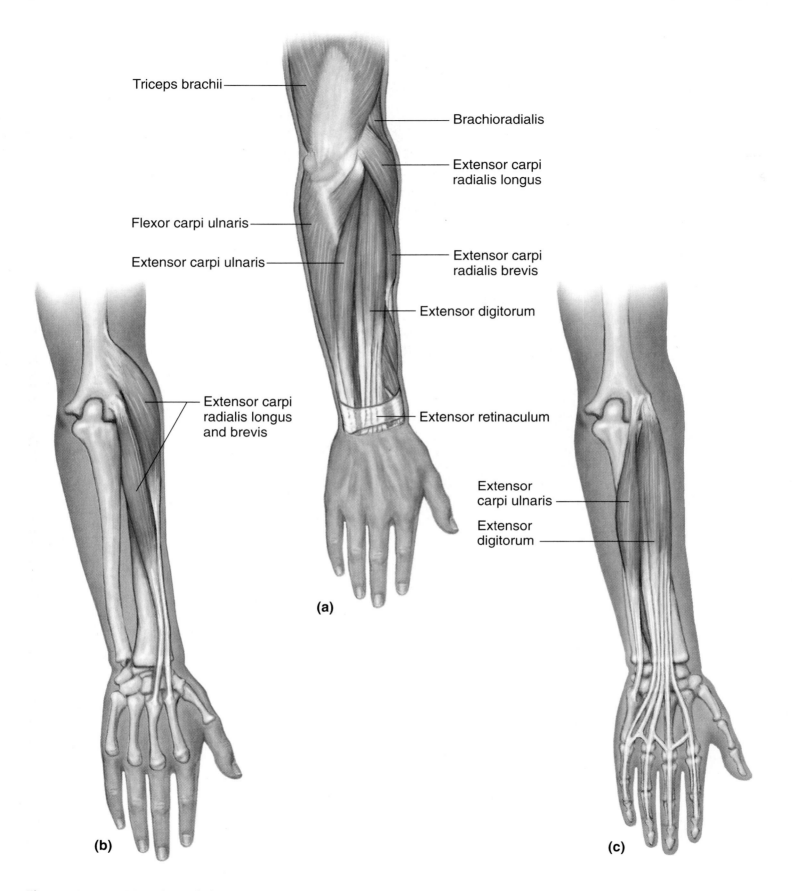

Triceps brachii

Brachioradialis

Extensor carpi
radialis longus

Flexor carpi ulnaris

Extensor carpi ulnaris

Extensor carpi
radialis brevis

Extensor digitorum

Extensor carpi
radialis longus
and brevis

Extensor retinaculum

Extensor
carpi ulnaris

Extensor
digitorum

(a)

(b)

(c)

Figure 9.30   Muscles of the arm and forearm

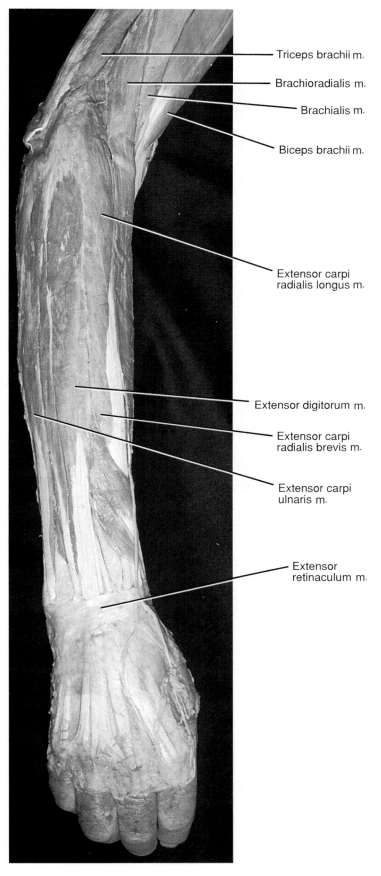

Triceps brachii m.

Brachioradialis m.

Brachialis m.

Biceps brachii m.

Extensor carpi
radialis longus m.

Extensor digitorum m.

Extensor carpi
radialis brevis m.

Extensor carpi
ulnaris m.

Extensor
retinaculum m.

Plate 65 Posterior view of the right forearm and hand

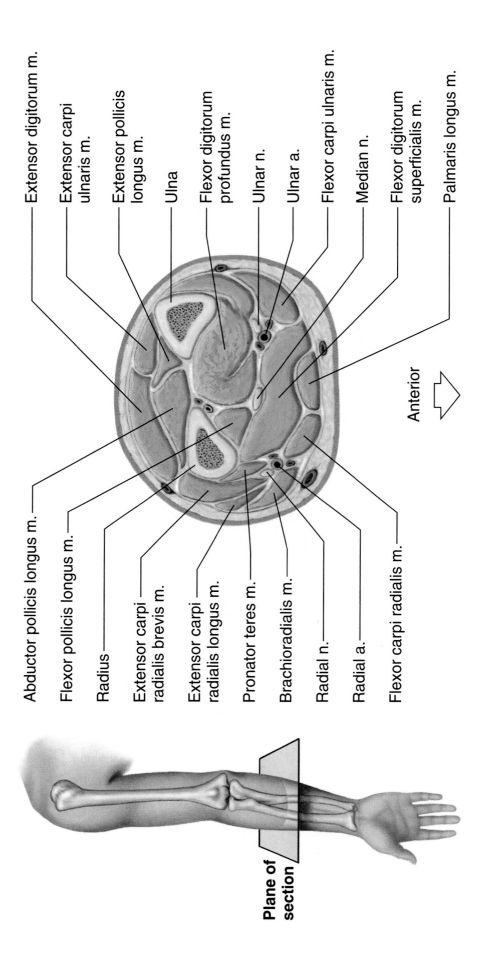

Extensor digitorum m.

Extensor carpi ulnaris m.

Extensor pollicis longus m.

Ulna

Flexor digitorum profundus m.

Ulnar n.

Ulnar a.

Flexor carpi ulnaris m.

Median n.

Flexor digitorum superficialis m.

Palmaris longus m.

Anterior

Abductor pollicis longus m.

Flexor pollicis longus m.

Radius

Extensor carpi radialis brevis m.

Extensor carpi radialis longus m.

Pronator teres m.

Brachioradialis m.

Radial n.

Radial a.

Flexor carpi radialis m.

Plane of section

Figure 9.31   A cross section of the forearm (superior view)

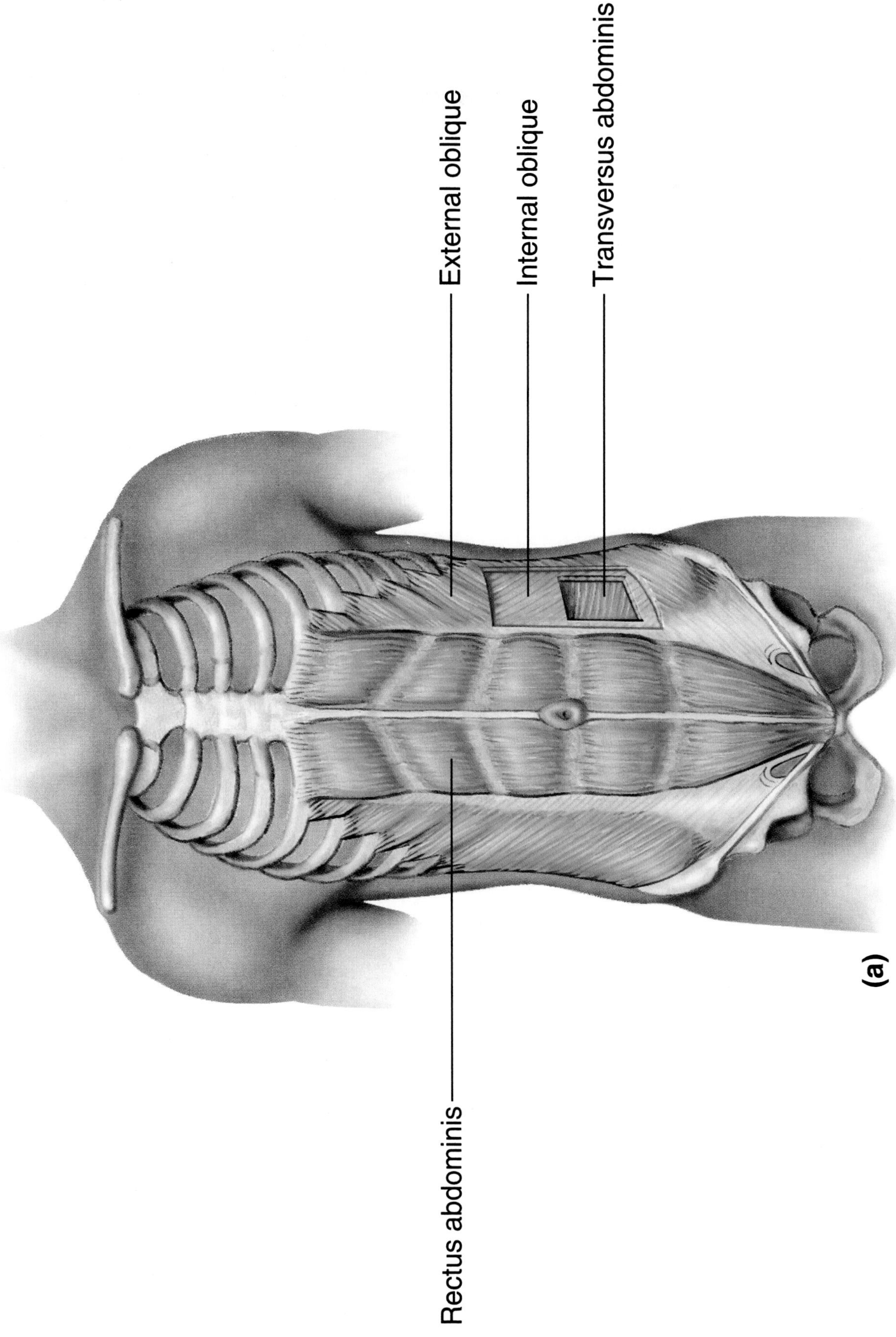

External oblique

Internal oblique

Transversus abdominis

Rectus abdominis

(a)

Figure 9.32a Isolated muscles of the abdominal wall

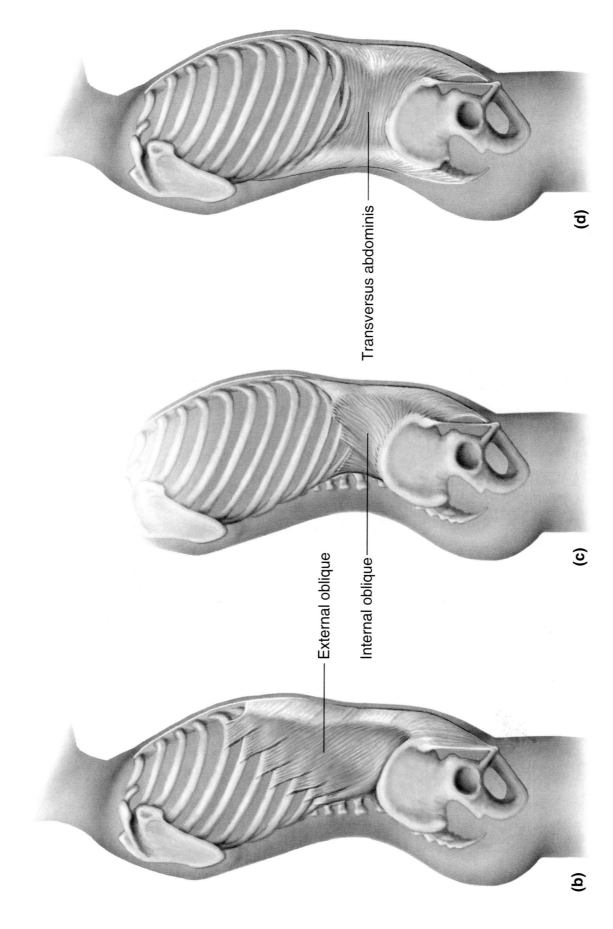

External oblique

Internal oblique

Transversus abdominis

(b)

(c)

(d)

Figure 9.32b–d Isolated muscles of the abdominal wall

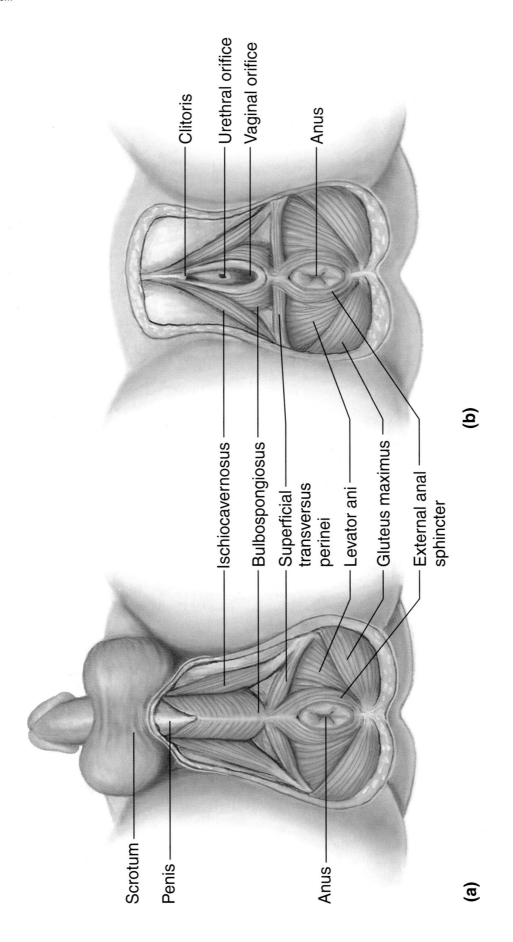

Scrotum

Penis

Anus

Ischiocavernosus

Bulbospongiosus

Superficial
transversus
perinei

Levator ani

Gluteus maximus

External anal
sphincter

Clitoris

Urethral orifice

Vaginal orifice

Anus

(a)

(b)

Figure 9.33a–b   External view of muscles of the male pelvic outlet and female pelvic outlet

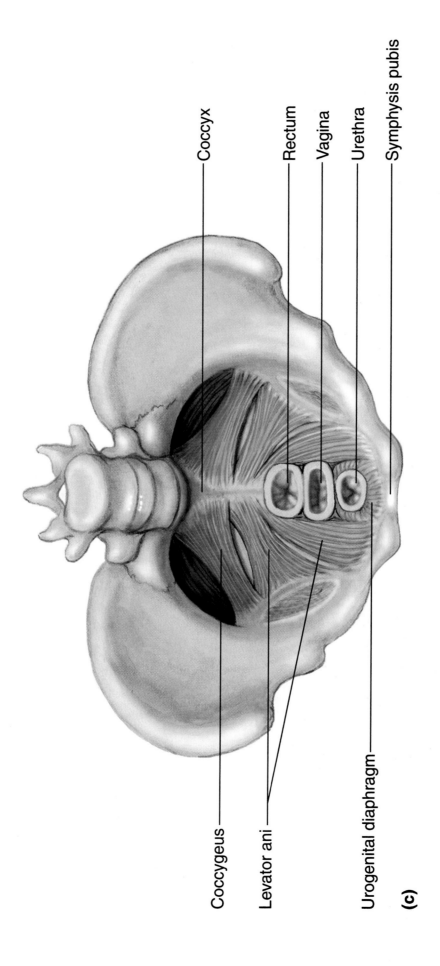

Coccyx

Rectum

Vagina

Urethra

Symphysis pubis

Coccygeus

Levator ani

Urogenital diaphragm

(c)

Figure 9.33c   Internal view of the female pelvic and urogenital diaphragms

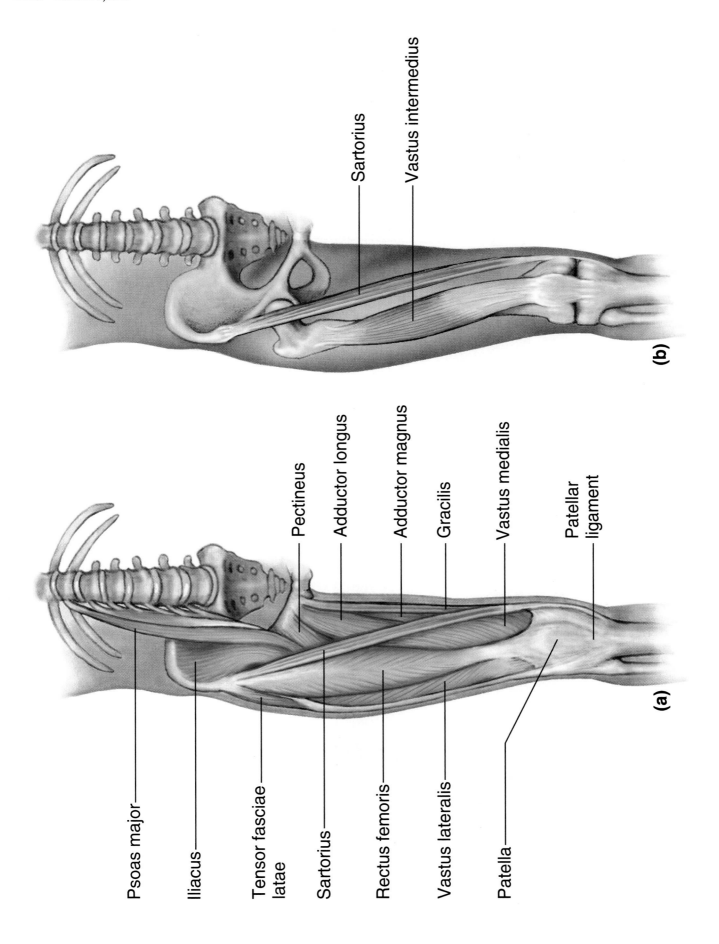

Figure 9.34a–b Muscles of the anterior right thigh and isolated view of the vastus intermedius

Sartorius

Vastus intermedius

(b)

Pectineus

Adductor longus

Adductor magnus

Gracilis

Vastus medialis

Patellar ligament

Psoas major

Iliacus

Tensor fasciae latae

Sartorius

Rectus femoris

Vastus lateralis

Patella

(a)

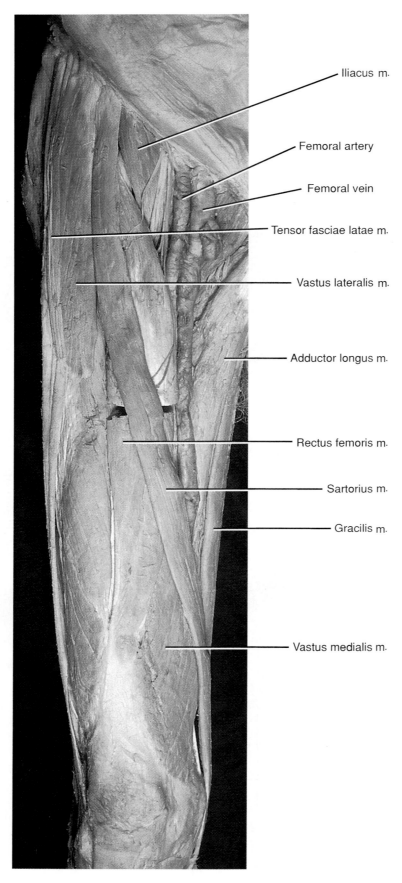

Iliacus m.

Femoral artery

Femoral vein

Tensor fasciae latae m.

Vastus lateralis m.

Adductor longus m.

Rectus femoris m.

Sartorius m.

Gracilis m.

Vastus medialis m.

Plate 66   Anterior view of the right thigh

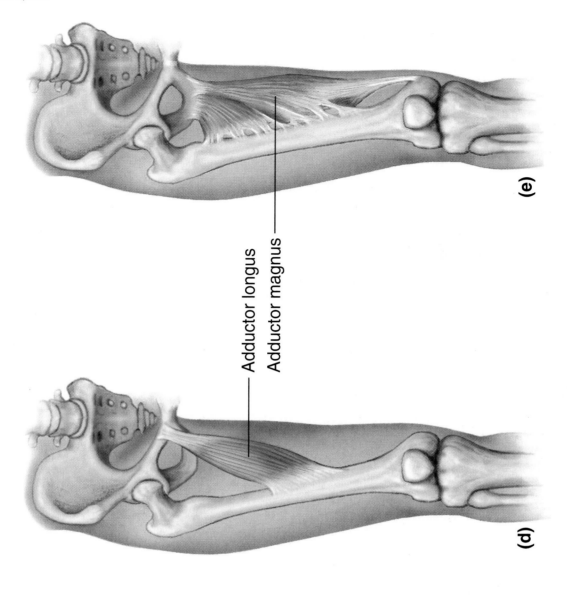

Adductor longus
Adductor magnus

Gracilis

(e)

(d)

(c)

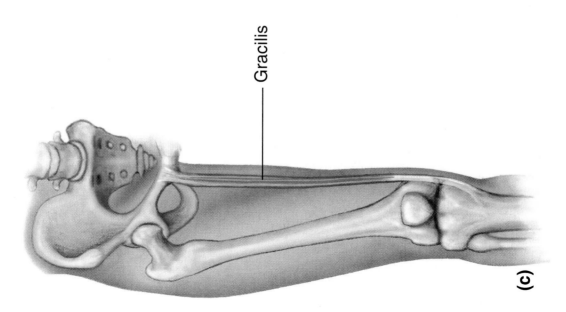

Figure 9.34c–e Adductors of the thigh

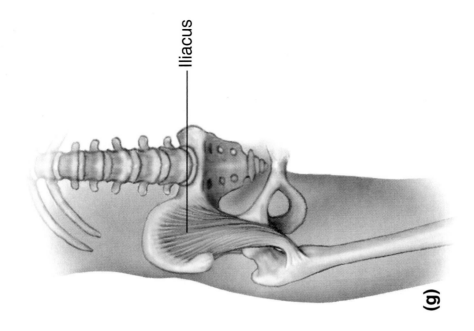

Iliacus

(g)

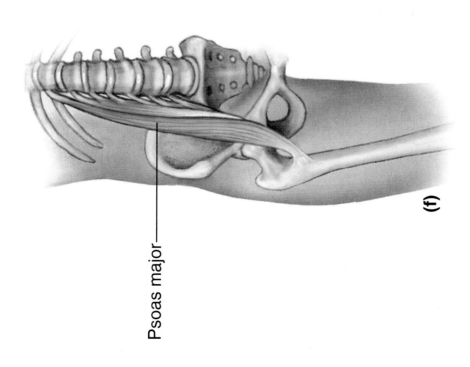

Psoas major

(f)

Figure 9.34f–g   Flexors of the thigh

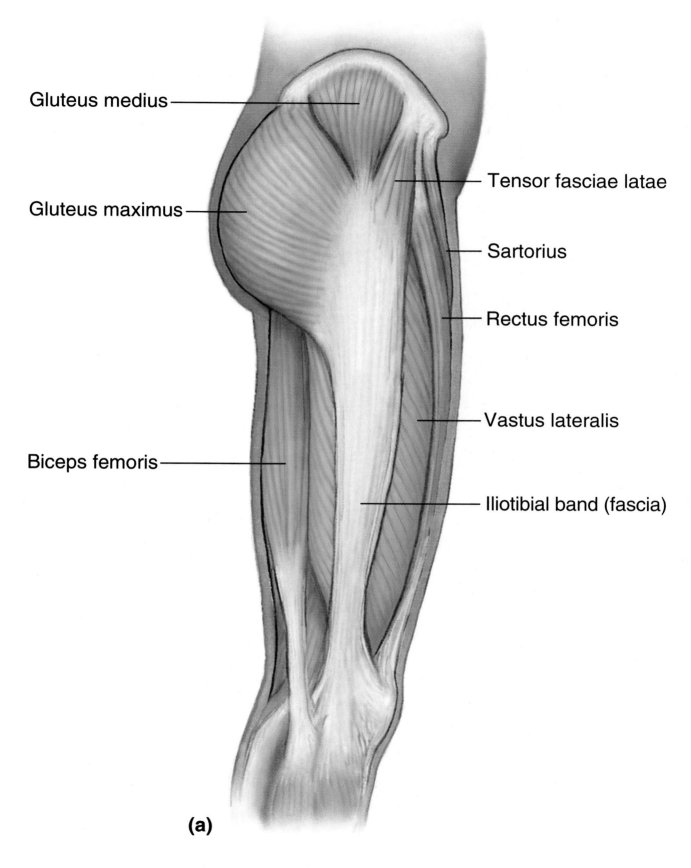

Gluteus medius

Gluteus maximus

Biceps femoris

Tensor fasciae latae

Sartorius

Rectus femoris

Vastus lateralis

Iliotibial band (fascia)

(a)

Figure 9.35a Muscles of the lateral right thigh

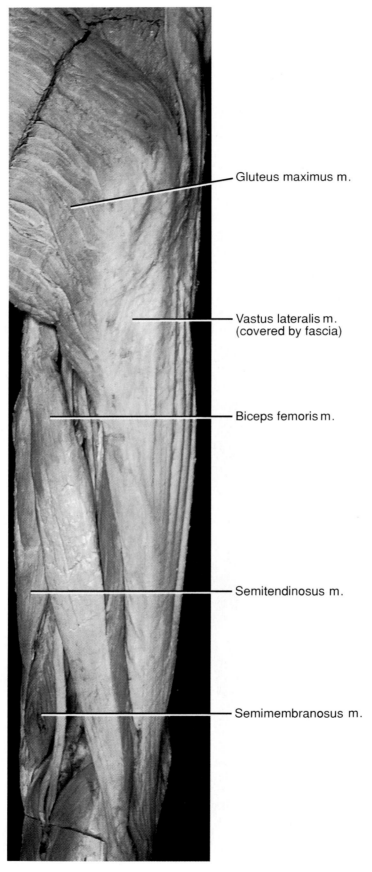

Gluteus maximus m.

Vastus lateralis m.
(covered by fascia)

Biceps femoris m.

Semitendinosus m.

Semimembranosus m.

Plate 67 Posterior view of the right thigh

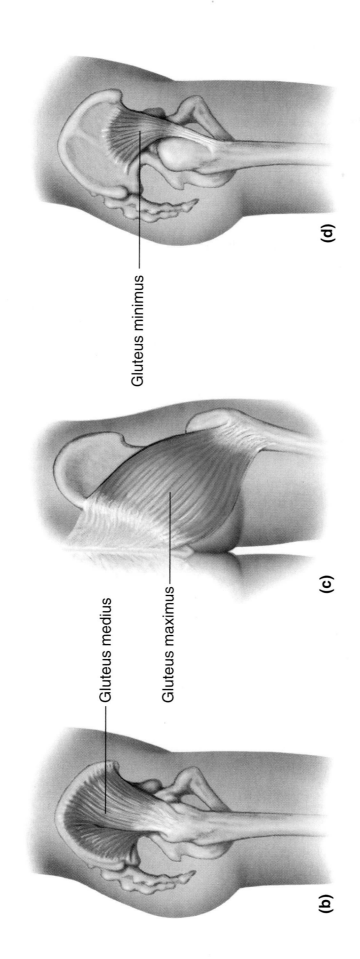

Figure 9.35b–d Isolated views of the gluteus muscles

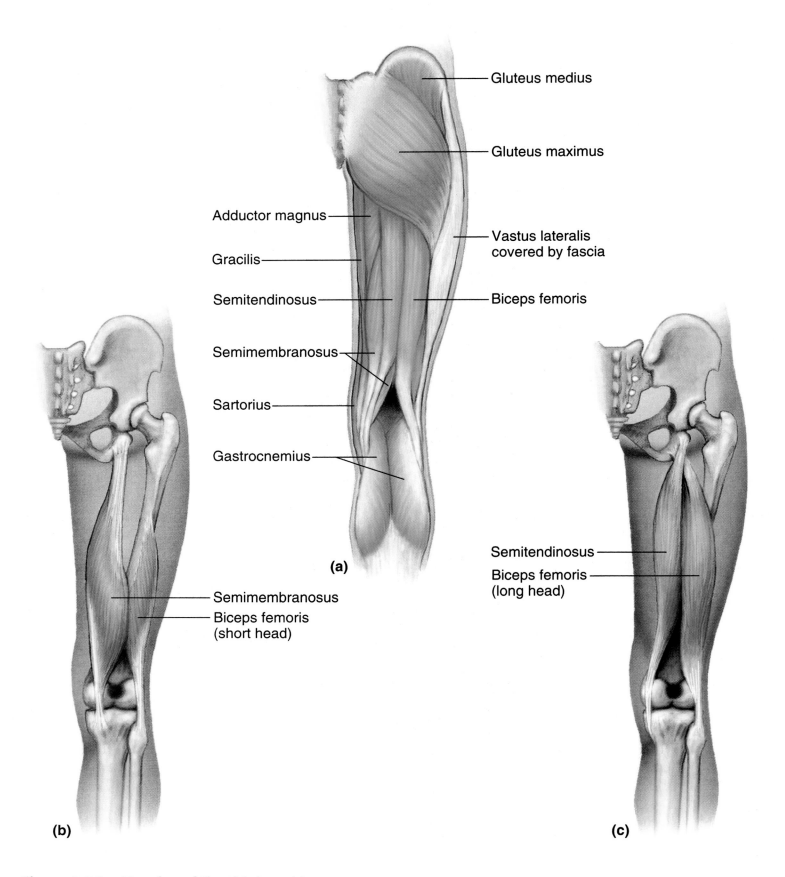

Gluteus medius

Gluteus maximus

Adductor magnus

Vastus lateralis
covered by fascia

Gracilis

Semitendinosus

Biceps femoris

Semimembranosus

Sartorius

Gastrocnemius

(a)

Semimembranosus

Biceps femoris
(short head)

(b)

Semitendinosus

Biceps femoris
(long head)

(c)

Figure 9.36   Muscles of the thigh and leg

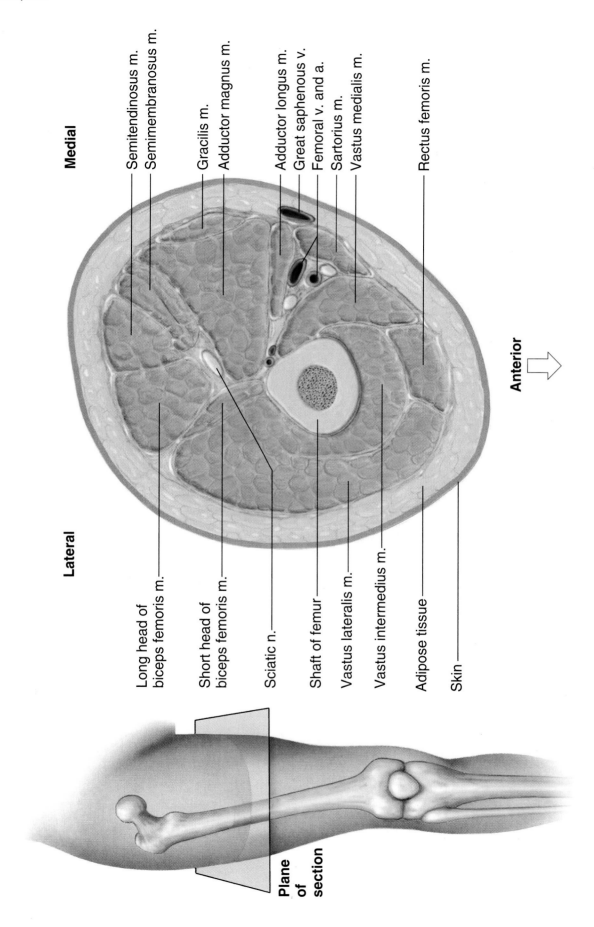

Medial

Semitendinosus m.
Semimembranosus m.
Gracilis m.
Adductor magnus m.
Adductor longus m.
Great saphenous v.
Femoral v. and a.
Sartorius m.
Vastus medialis m.
Rectus femoris m.

Lateral

Long head of biceps femoris m.
Short head of biceps femoris m.
Sciatic n.
Shaft of femur
Vastus lateralis m.
Vastus intermedius m.
Adipose tissue
Skin

Anterior

Plane of section

Figure 9.37   A cross section of the thigh

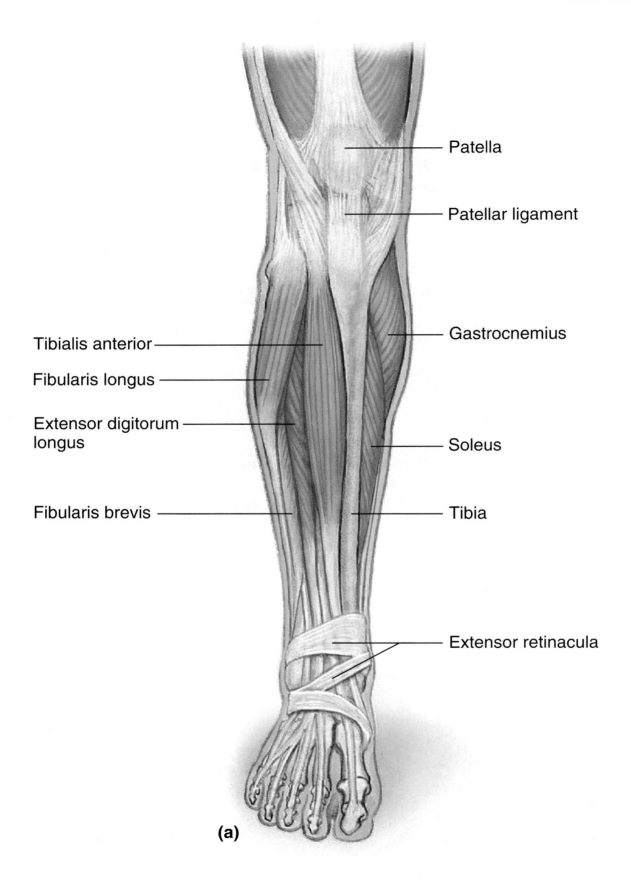

Patella

Patellar ligament

Gastrocnemius

Tibialis anterior

Fibularis longus

Extensor digitorum longus

Soleus

Fibularis brevis

Tibia

Extensor retinacula

(a)

Figure 9.38a   Muscles of the anterior right leg

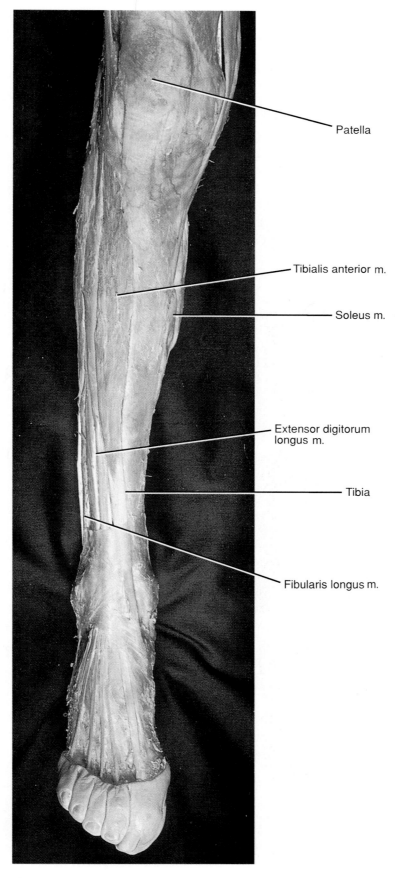

Patella

Tibialis anterior m.

Soleus m.

Extensor digitorum longus m.

Tibia

Fibularis longus m.

Plate 68   Anterior view of the right leg

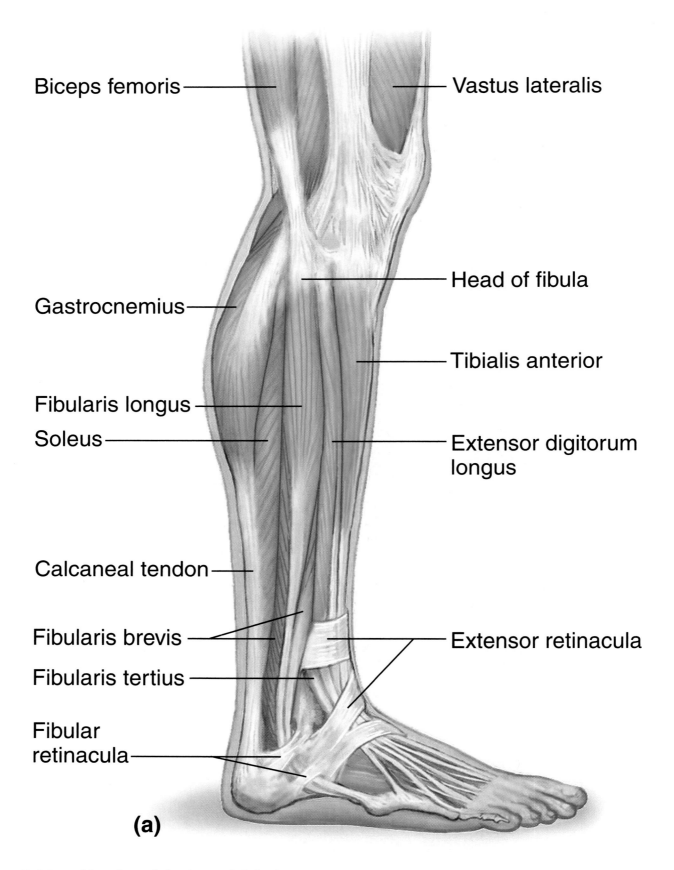

Biceps femoris

Vastus lateralis

Gastrocnemius

Head of fibula

Tibialis anterior

Fibularis longus

Soleus

Extensor digitorum
longus

Calcaneal tendon

Fibularis brevis

Fibularis tertius

Extensor retinacula

Fibular
retinacula

(a)

Figure 9.39a   Muscles of the lateral right leg

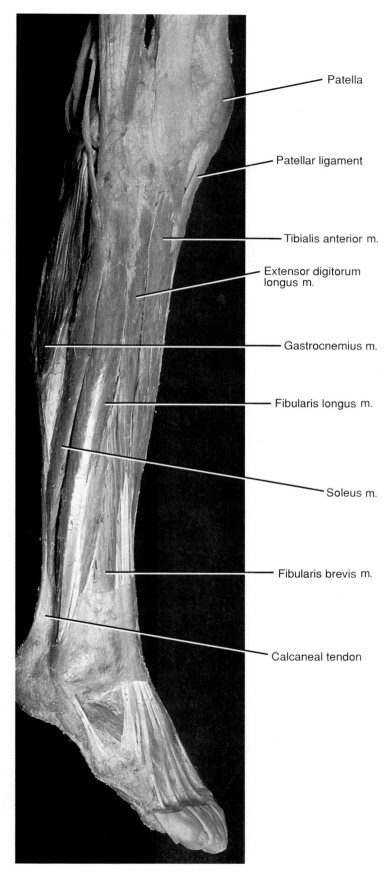

Patella

Patellar ligament

Tibialis anterior m.

Extensor digitorum
longus m.

Gastrocnemius m.

Fibularis longus m.

Soleus m.

Fibularis brevis m.

Calcaneal tendon

Plate 69 **Lateral view of the right leg**

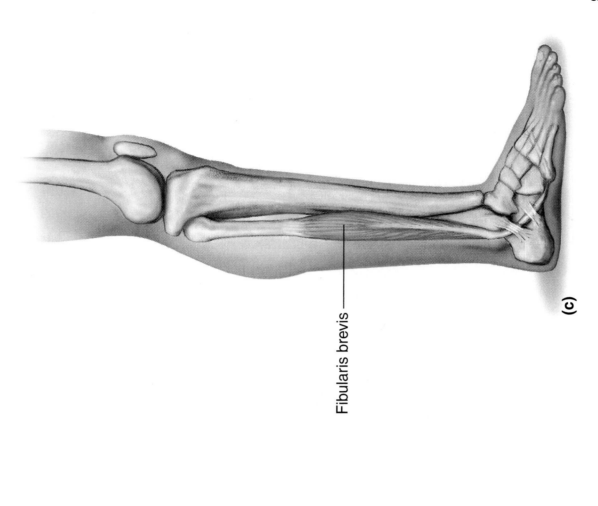

Fibularis brevis

(c)

Fibularis longus

(b)

Figure 9.39b–c   Isolated views of fibularis longus and fibularis brevis

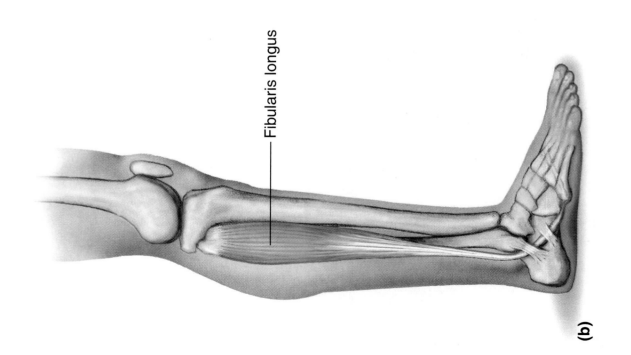

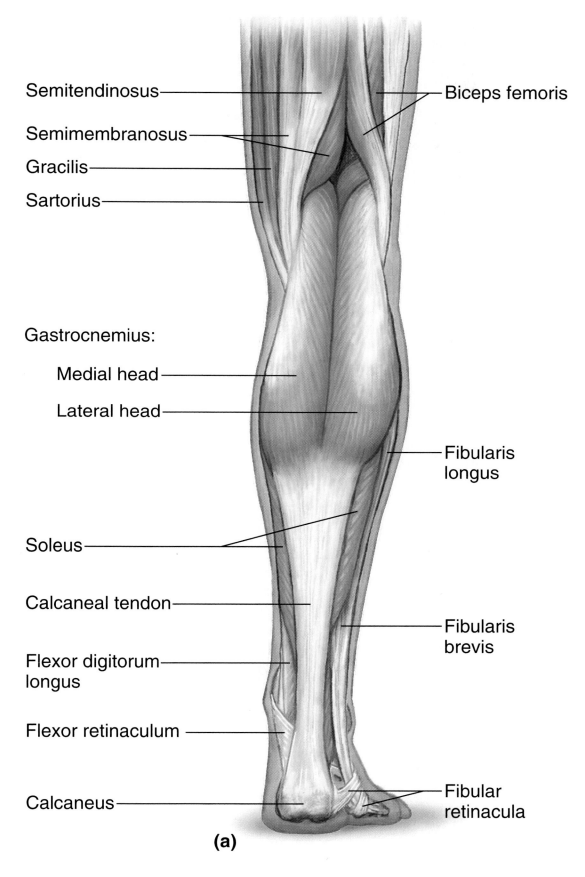

Semitendinosus

Semimembranosus

Gracilis

Sartorius

Biceps femoris

Gastrocnemius:

Medial head

Lateral head

Fibularis longus

Soleus

Calcaneal tendon

Fibularis brevis

Flexor digitorum longus

Flexor retinaculum

Calcaneus

Fibular retinacula

(a)

Figure 9.40a   Muscles of the posterior right leg

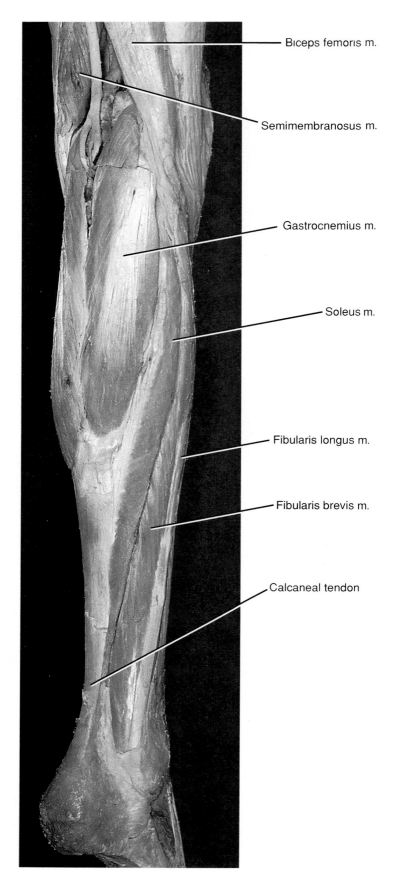

Biceps femoris m.

Semimembranosus m.

Gastrocnemius m.

Soleus m.

Fibularis longus m.

Fibularis brevis m.

Calcaneal tendon

Plate 70   Posterior view of the right leg

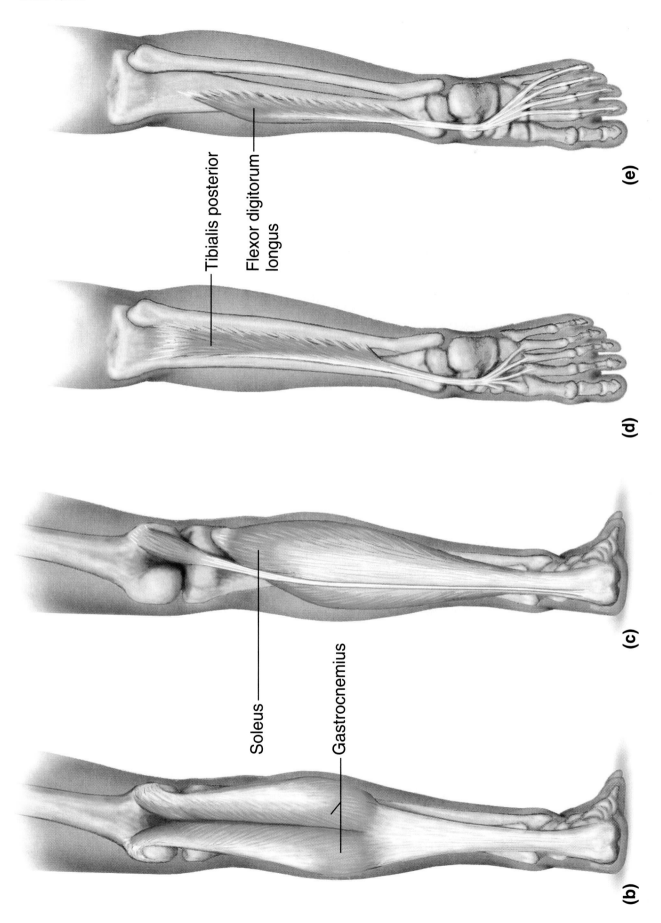

Figure 9.40b–e   Isolated views of muscles associated with the posterior right leg

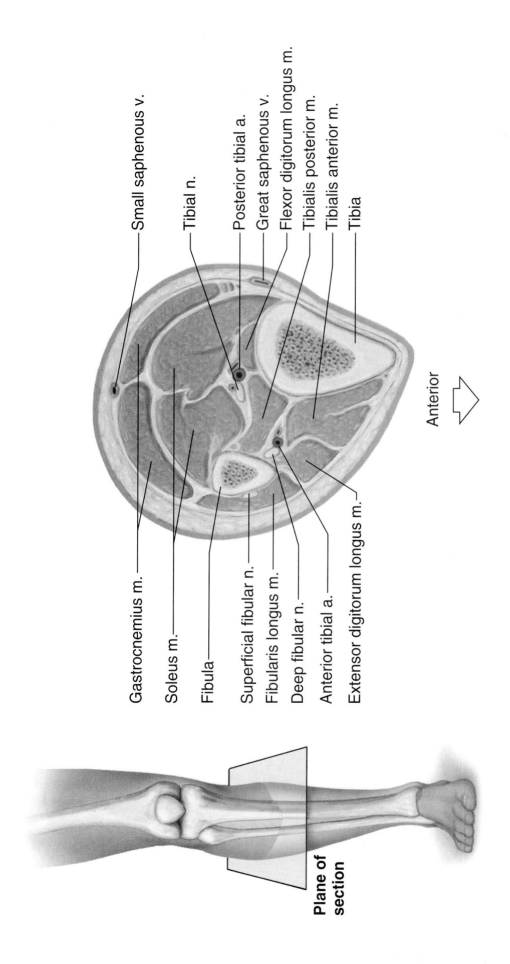

Small saphenous v.

Tibial n.

Posterior tibial a.

Great saphenous v.

Flexor digitorum longus m.

Tibialis posterior m.

Tibialis anterior m.

Tibia

Anterior

Gastrocnemius m.

Soleus m.

Fibula

Superficial fibular n.

Fibularis longus m.

Deep fibular n.

Anterior tibial a.

Extensor digitorum longus m.

Plane of section

Figure 9.41   A cross section of the leg

Unit 4 Nervous System

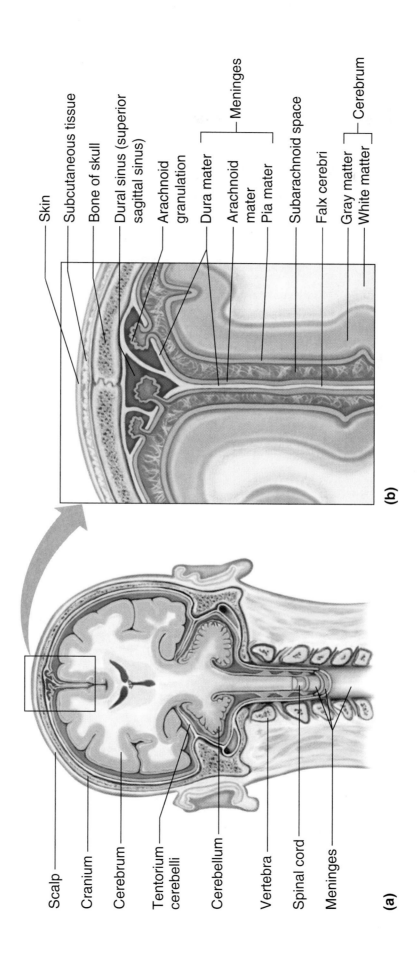

Skin

Subcutaneous tissue

Bone of skull

Dural sinus (superior sagittal sinus)

Arachnoid granulation

Dura mater ⎫
Arachnoid mater ⎬ Meninges
Pia mater ⎭

Subarachnoid space

Falx cerebri

Gray matter ⎫ Cerebrum
White matter ⎭

(b)

Scalp

Cranium

Cerebrum

Tentorium cerebelli

Cerebellum

Vertebra

Spinal cord

Meninges

(a)

Figure 11.1   Meninges

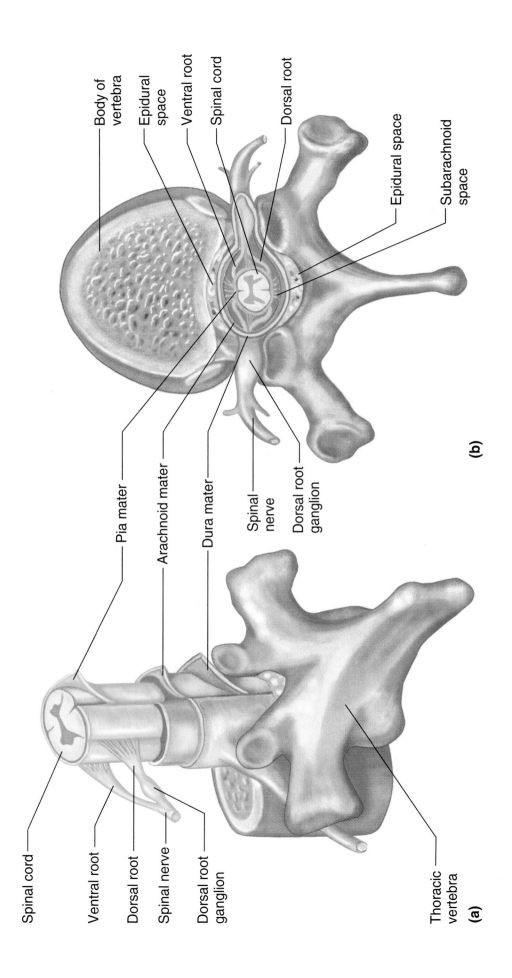

Body of vertebra

Epidural space

Ventral root

Spinal cord

Dorsal root

Epidural space

Subarachnoid space

(b)

Pia mater

Arachnoid mater

Dura mater

Spinal nerve

Dorsal root ganglion

Spinal cord

Ventral root

Dorsal root

Spinal nerve

Dorsal root ganglion

Thoracic vertebra

(a)

Figure 11.2   Meninges of the spinal cord

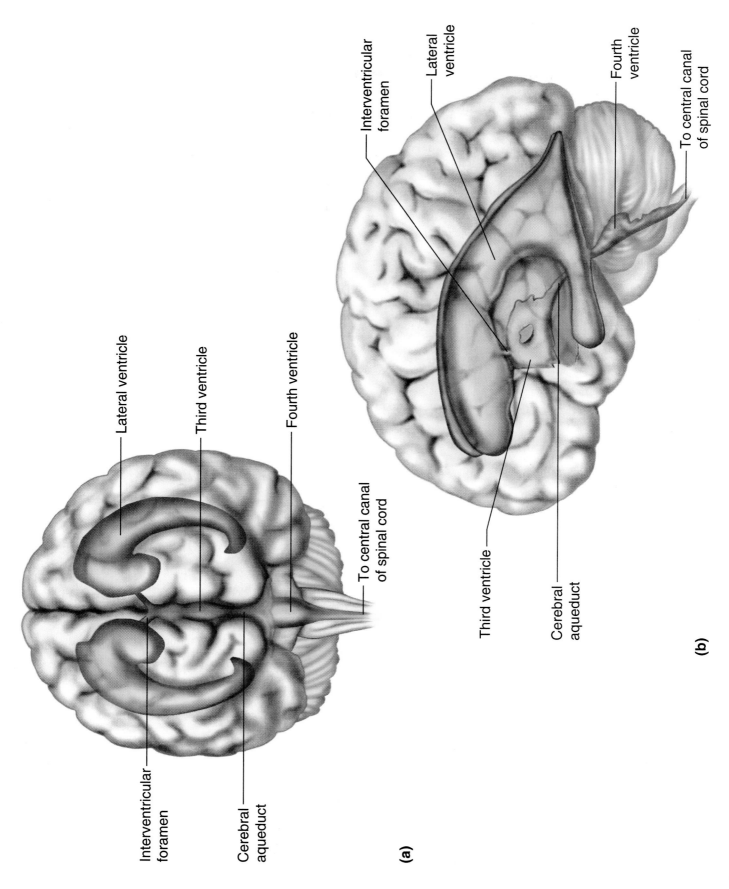

Lateral ventricle

Interventricular foramen

Lateral ventricle

Third ventricle

Fourth ventricle

Fourth ventricle

To central canal of spinal cord

Third ventricle

Cerebral aqueduct

To central canal of spinal cord

Interventricular foramen

Cerebral aqueduct

(a)

(b)

Figure 11.3 Ventricles in the brain

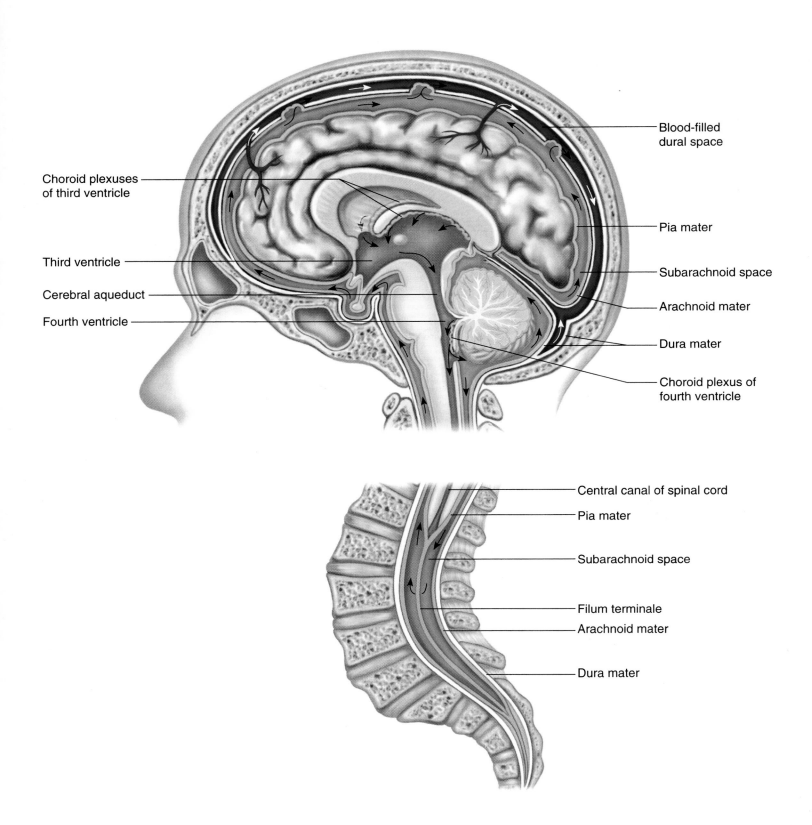

Choroid plexuses
of third ventricle

Third ventricle

Cerebral aqueduct

Fourth ventricle

Blood-filled
dural space

Pia mater

Subarachnoid space

Arachnoid mater

Dura mater

Choroid plexus of
fourth ventricle

Central canal of spinal cord

Pia mater

Subarachnoid space

Filum terminale

Arachnoid mater

Dura mater

Figure 11.4   Choroid plexuses in ventricle walls secrete cerebrospinal fluid

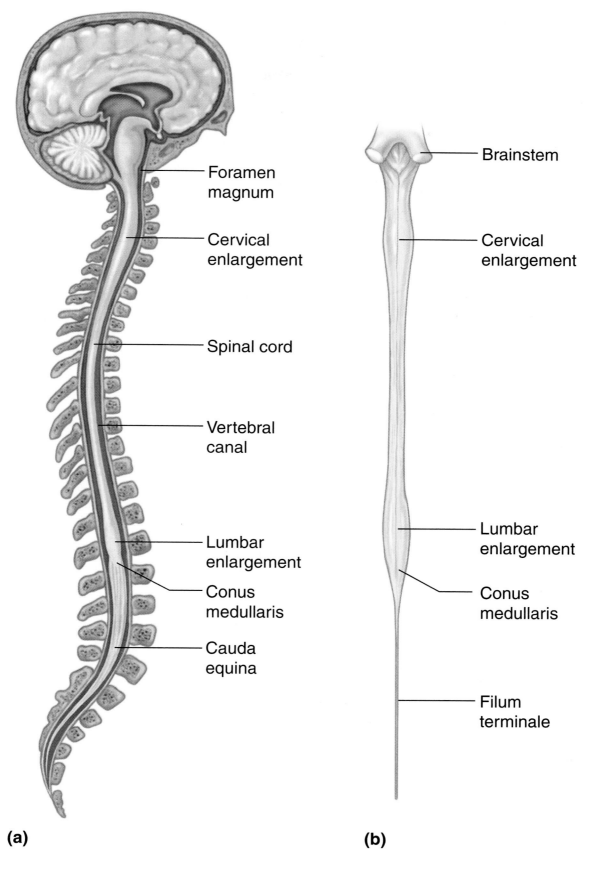

(a)

Foramen magnum

Cervical enlargement

Spinal cord

Vertebral canal

Lumbar enlargement

Conus medullaris

Cauda equina

(b)

Brainstem

Cervical enlargement

Lumbar enlargement

Conus medullaris

Filum terminale

Figure 11.5   Spinal cord

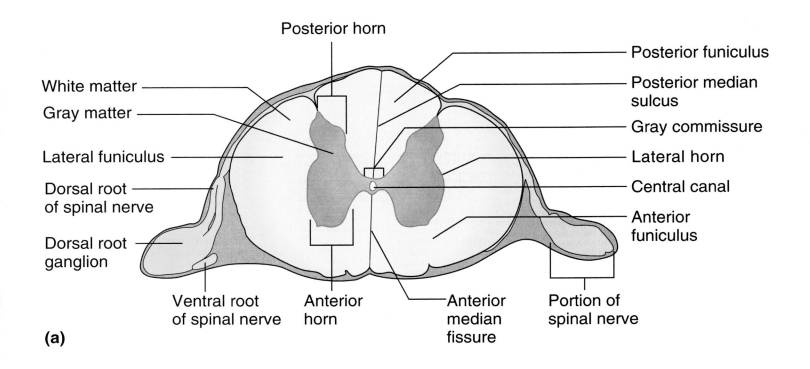

(a)

Posterior horn

Posterior funiculus

Posterior median sulcus

White matter

Gray matter

Gray commissure

Lateral funiculus

Lateral horn

Dorsal root of spinal nerve

Central canal

Dorsal root ganglion

Anterior funiculus

Ventral root of spinal nerve

Anterior horn

Anterior median fissure

Portion of spinal nerve

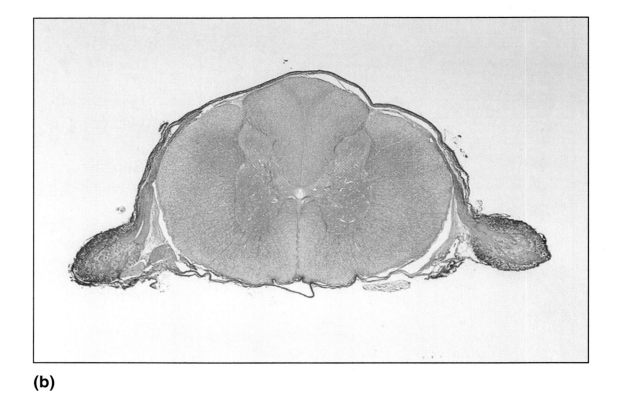

(b)

Figure 11.6   Spinal cord

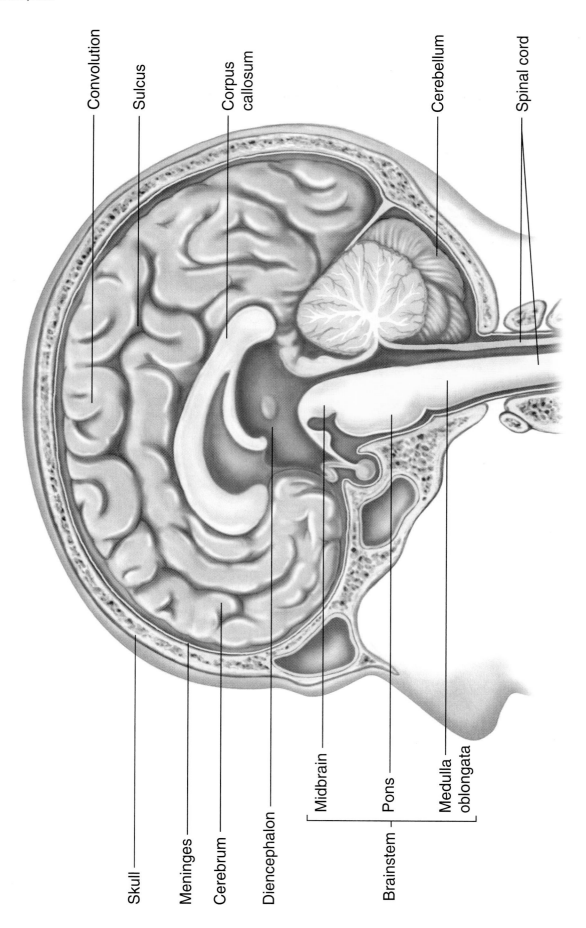

Convolution

Sulcus

Corpus callosum

Cerebellum

Spinal cord

Skull

Meninges

Cerebrum

Diencephalon

Midbrain

Pons

Medulla oblongata

Brainstem

Figure 11.15 **The major portions of the brain**

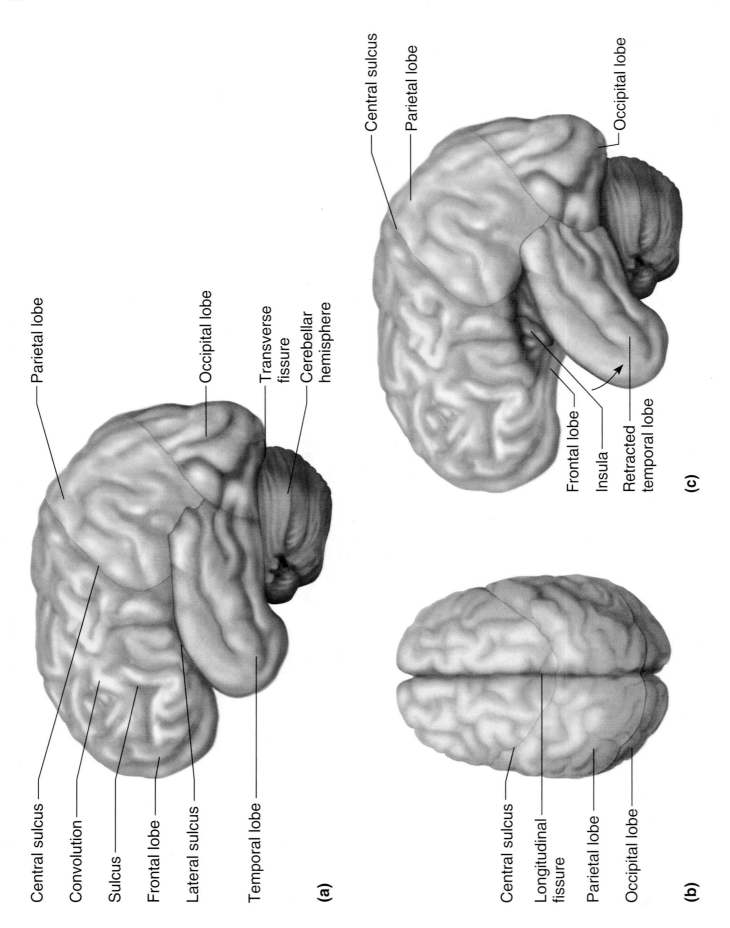

(a)

Parietal lobe

Occipital lobe

Transverse fissure

Cerebellar hemisphere

Central sulcus

Convolution

Sulcus

Frontal lobe

Lateral sulcus

Temporal lobe

(c)

Central sulcus

Parietal lobe

Occipital lobe

Frontal lobe

Insula

Retracted temporal lobe

(b)

Central sulcus

Longitudinal fissure

Parietal lobe

Occipital lobe

Figure 11.16   Colors in this figure distinguish the lobes of the cerebral hemispheres

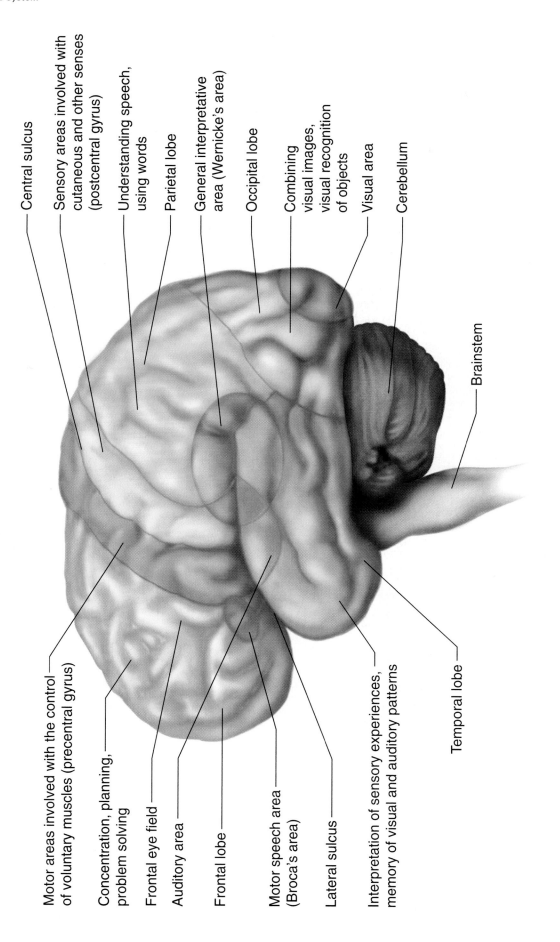

Central sulcus

Sensory areas involved with cutaneous and other senses (postcentral gyrus)

Understanding speech, using words

Parietal lobe

General interpretative area (Wernicke's area)

Occipital lobe

Combining visual images, visual recognition of objects

Visual area

Cerebellum

Brainstem

Motor areas involved with the control of voluntary muscles (precentral gyrus)

Concentration, planning, problem solving

Frontal eye field

Auditory area

Frontal lobe

Motor speech area (Broca's area)

Lateral sulcus

Interpretation of sensory experiences, memory of visual and auditory patterns

Temporal lobe

Figure 11.17 Some motor, sensory, and association areas of the left cerebral cortex

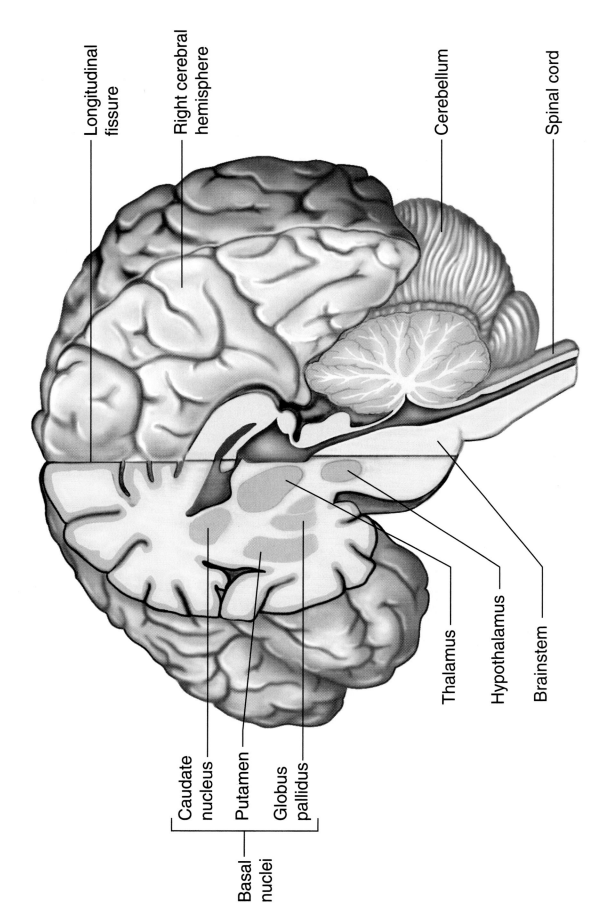

Longitudinal fissure

Right cerebral hemisphere

Cerebellum

Spinal cord

Caudate nucleus

Putamen

Globus pallidus

Basal nuclei

Thalamus

Hypothalamus

Brainstem

Figure 11.19   **A coronal section of the left cerebral hemisphere reveals some of the basal nuclei**

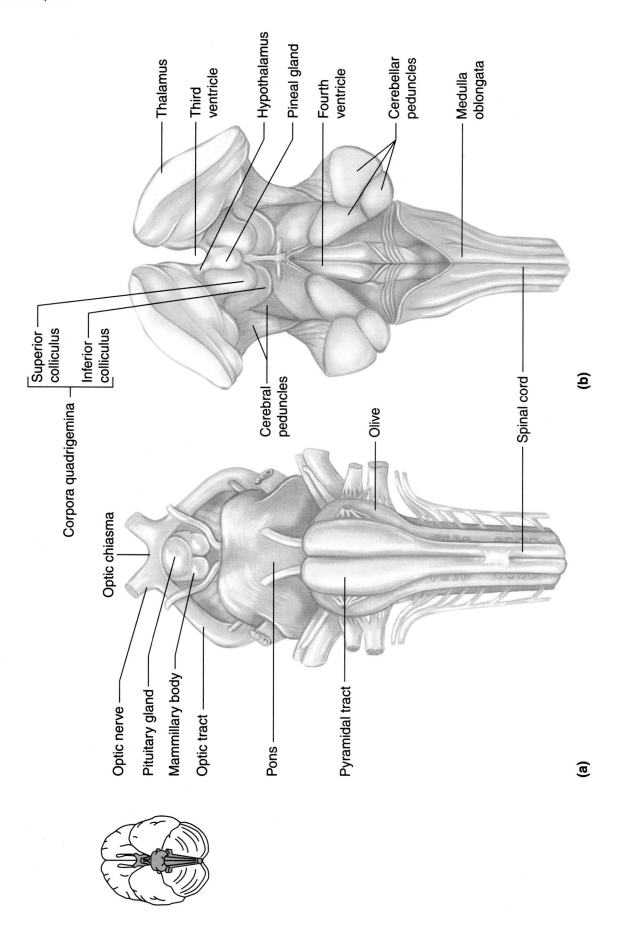

Thalamus

Third
ventricle

Hypothalamus

Pineal gland

Fourth
ventricle

Cerebellar
peduncles

Medulla
oblongata

Superior
colliculus

Inferior
colliculus

Corpora quadrigemina

Cerebral
peduncles

Optic chiasma

Optic nerve

Pituitary gland

Mammillary body

Optic tract

Pons

Pyramidal tract

Olive

Spinal cord

(a)

(b)

Figure 11.20 Brainstem

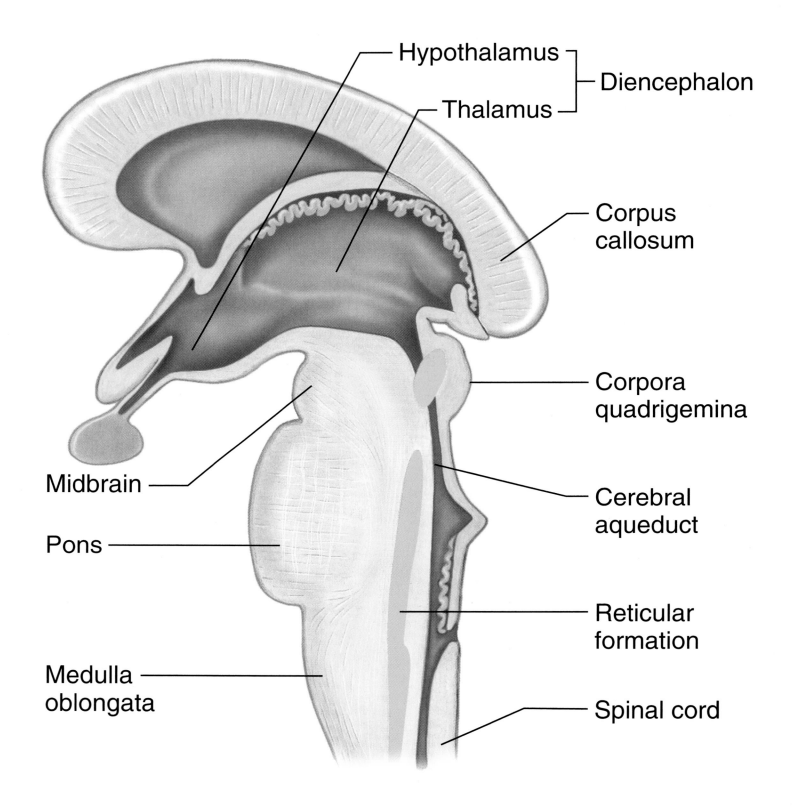

Figure 11.21   The reticular formation

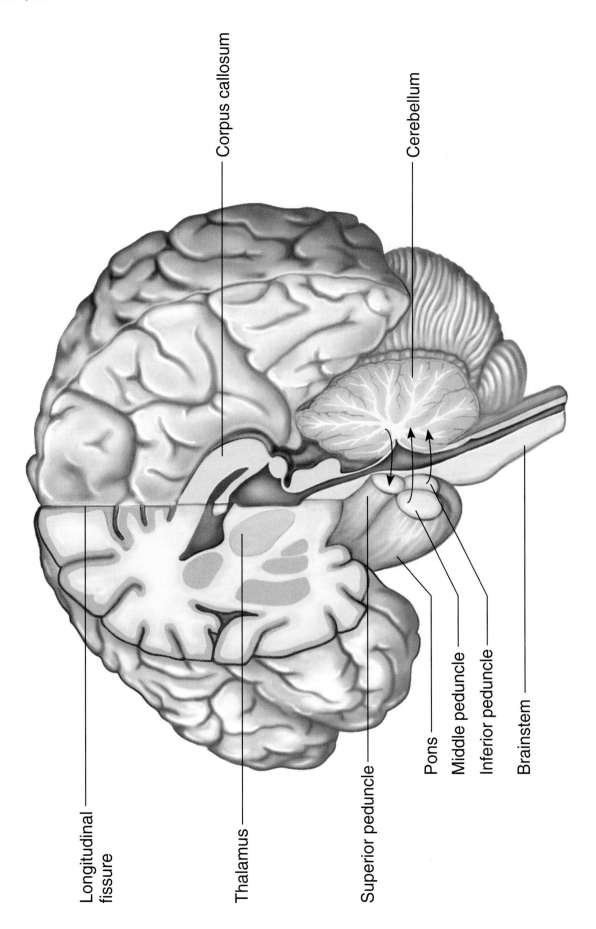

Corpus callosum

Cerebellum

Longitudinal
fissure

Thalamus

Superior peduncle

Pons

Middle peduncle

Inferior peduncle

Brainstem

Figure 11.22 The cerebellum communicates with other parts of the nervous system

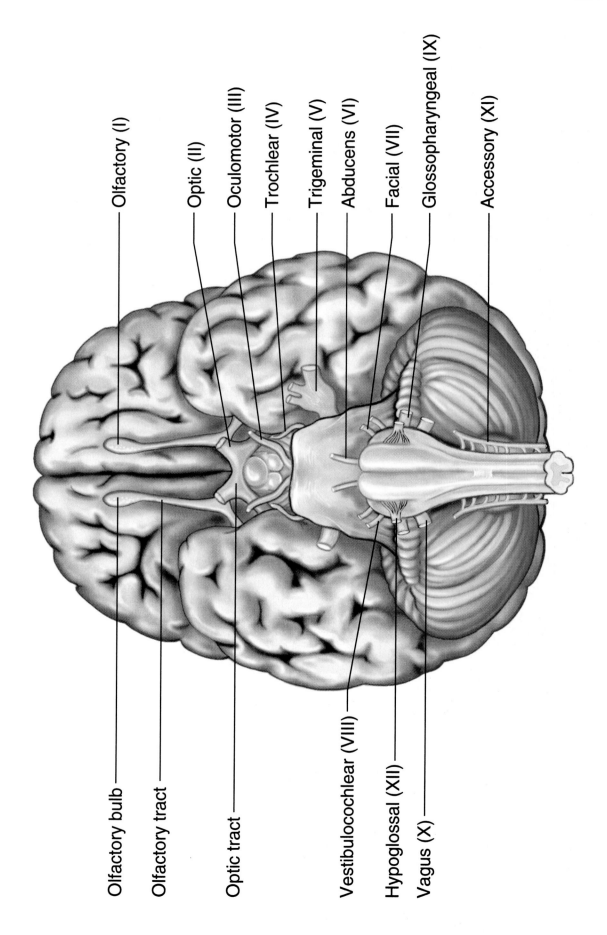

Olfactory (I)

Optic (II)

Oculomotor (III)

Trochlear (IV)

Trigeminal (V)

Abducens (VI)

Facial (VII)

Glossopharyngeal (IX)

Accessory (XI)

Olfactory bulb

Olfactory tract

Optic tract

Vestibulocochlear (VIII)

Hypoglossal (XII)

Vagus (X)

Figure 11.25   The cranial nerves, except for the first pair, arise from the brainstem

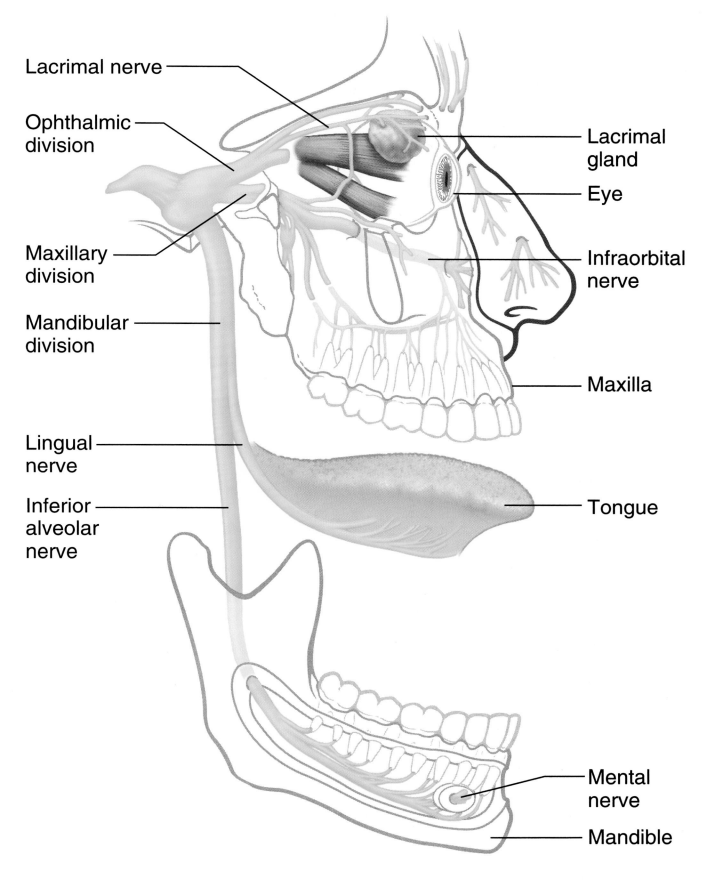

Figure 11.26 Each trigeminal nerve has three large branches

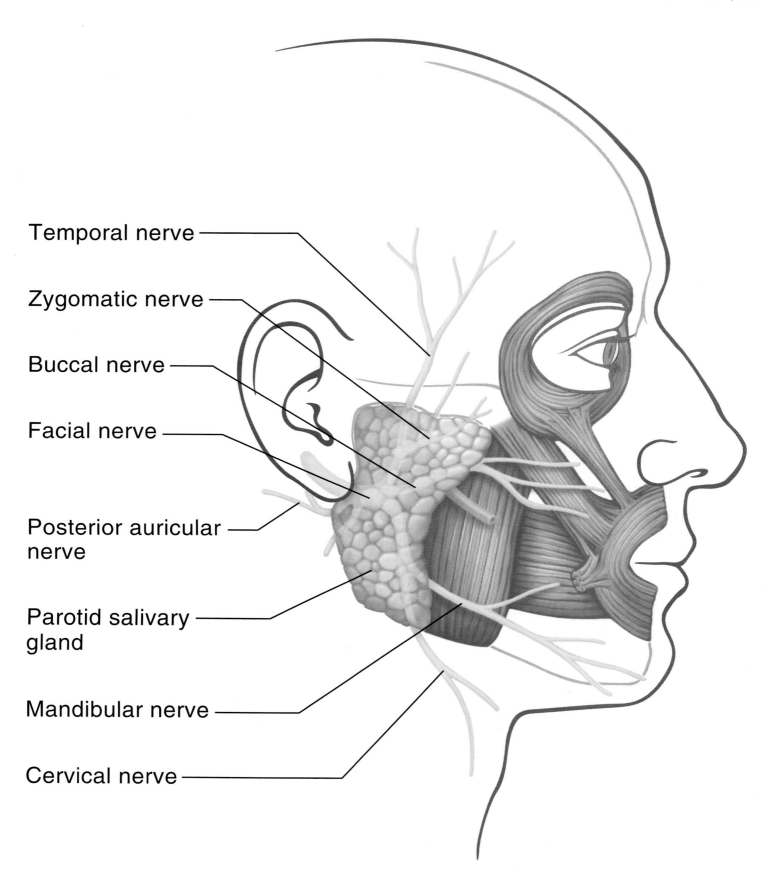

Temporal nerve

Zygomatic nerve

Buccal nerve

Facial nerve

Posterior auricular
nerve

Parotid salivary
gland

Mandibular nerve

Cervical nerve

Figure 11.27   The facial nerves

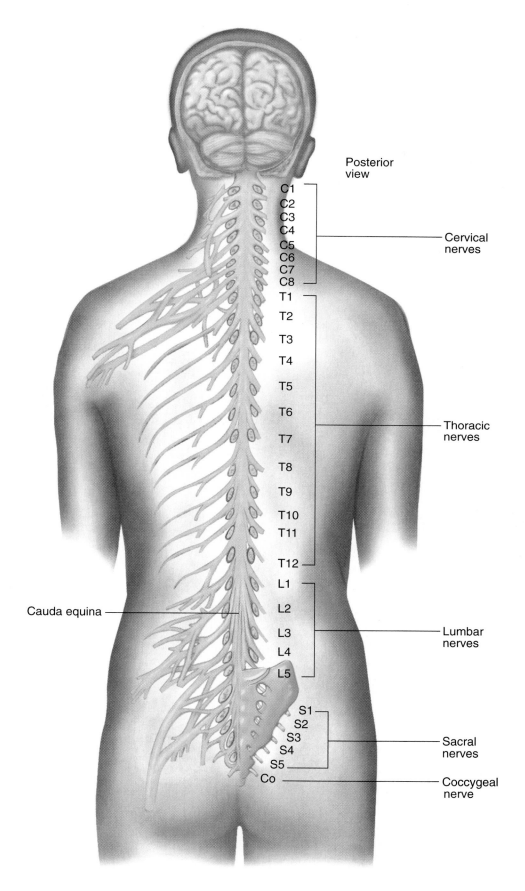

Posterior view

C1
C2
C3
C4
C5
C6
C7
C8 — Cervical nerves

T1
T2
T3
T4
T5
T6
T7 — Thoracic nerves
T8
T9
T10
T11
T12

L1
L2
L3 — Lumbar nerves
L4
L5

Cauda equina

S1
S2
S3 — Sacral nerves
S4
S5

Co — Coccygeal nerve

Figure 11.29 The thirty-one pairs of spinal nerves

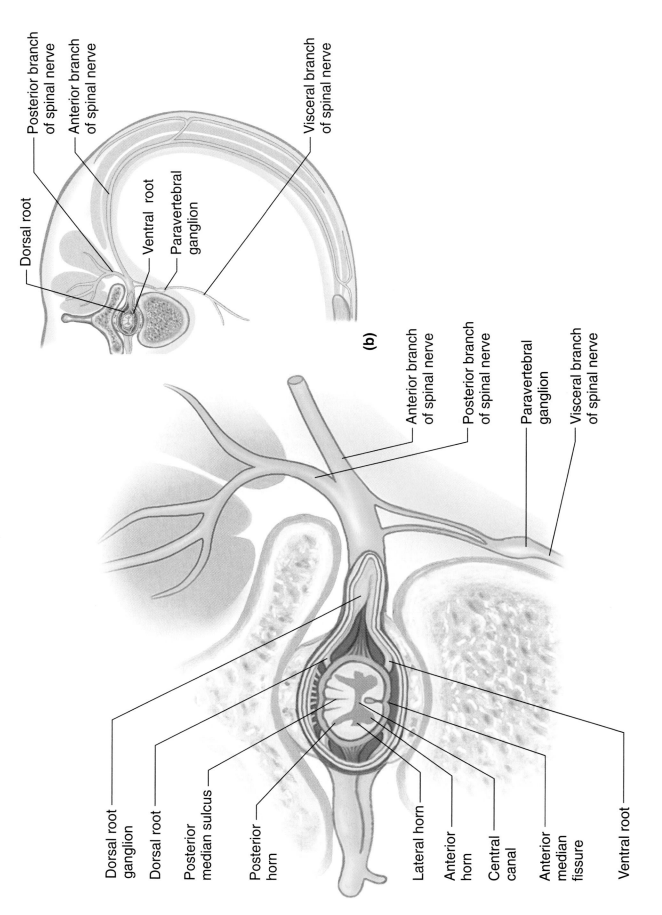

Posterior branch of spinal nerve

Anterior branch of spinal nerve

Dorsal root

Ventral root

Paravertebral ganglion

Visceral branch of spinal nerve

(b)

Anterior branch of spinal nerve

Posterior branch of spinal nerve

Paravertebral ganglion

Visceral branch of spinal nerve

Dorsal root ganglion

Dorsal root

Posterior median sulcus

Posterior horn

Lateral horn

Anterior horn

Central canal

Anterior median fissure

Ventral root

(a)

Figure 11.31   Spinal nerve

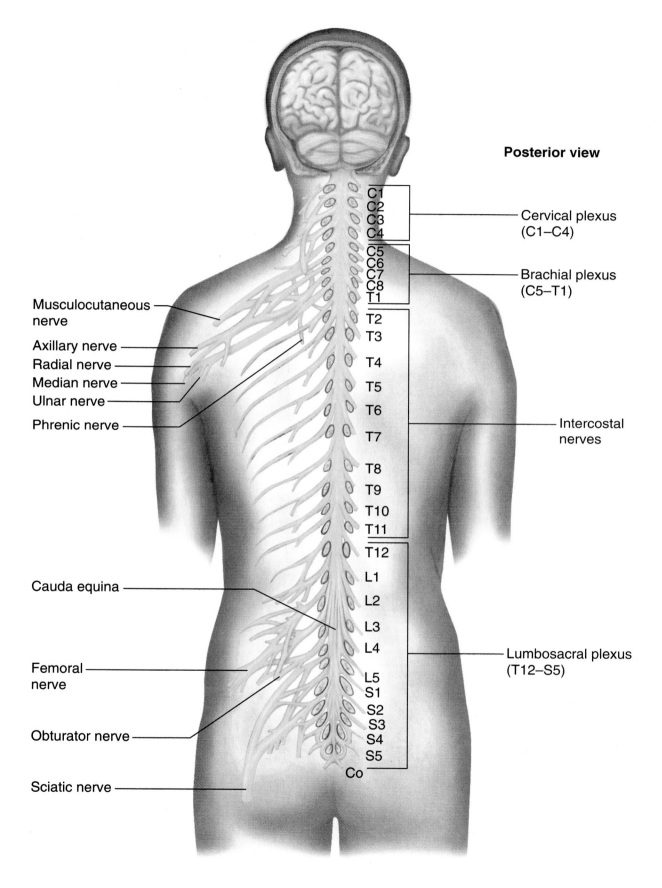

Posterior view

C1
C2
C3
C4
— Cervical plexus
(C1–C4)

C5
C6
C7
C8
T1
— Brachial plexus
(C5–T1)

Musculocutaneous nerve

Axillary nerve

Radial nerve

Median nerve

Ulnar nerve

Phrenic nerve

T2
T3
T4
T5
T6
T7
T8
T9
T10
T11

— Intercostal nerves

Cauda equina

T12
L1
L2
L3
L4
L5

Femoral nerve

S1
S2
S3
S4
S5
Co

— Lumbosacral plexus
(T12–S5)

Obturator nerve

Sciatic nerve

Figure 11.32   The anterior branches of the spinal nerves in the thoracic region

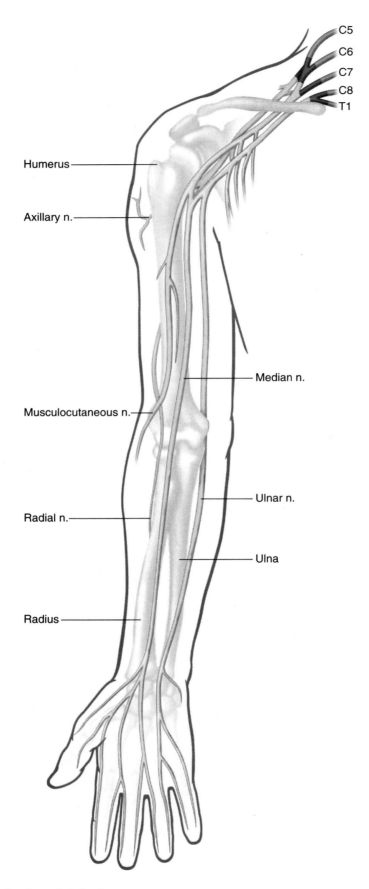

Figure 11.33   Nerves of the brachial plexus

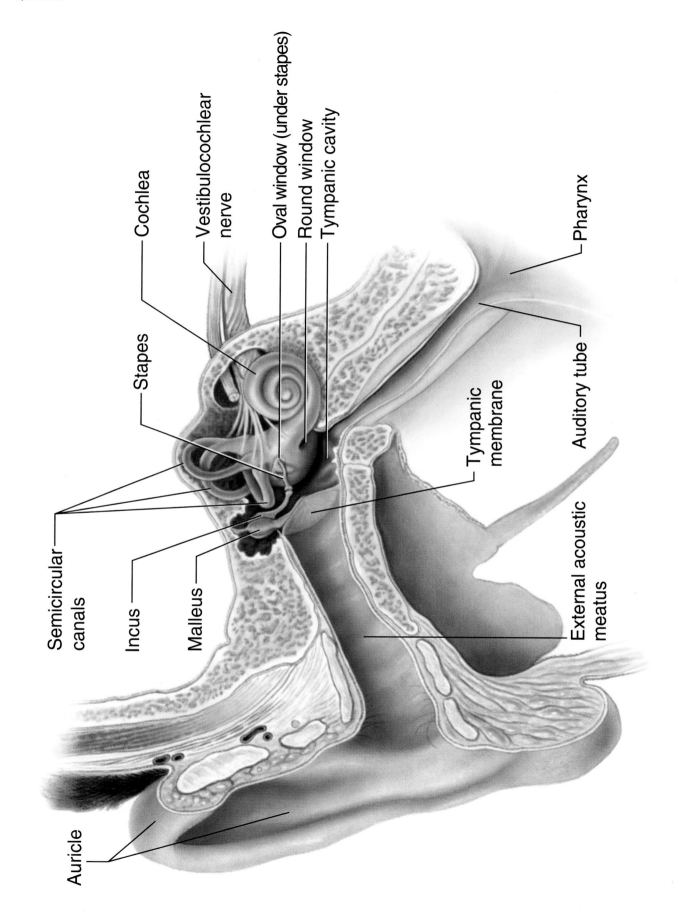

Cochlea

Vestibulocochlear nerve

Oval window (under stapes)

Round window

Tympanic cavity

Pharynx

Stapes

Tympanic membrane

Auditory tube

Semicircular canals

Incus

Malleus

External acoustic meatus

Auricle

Figure 12.9 Major parts of the ear

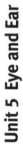

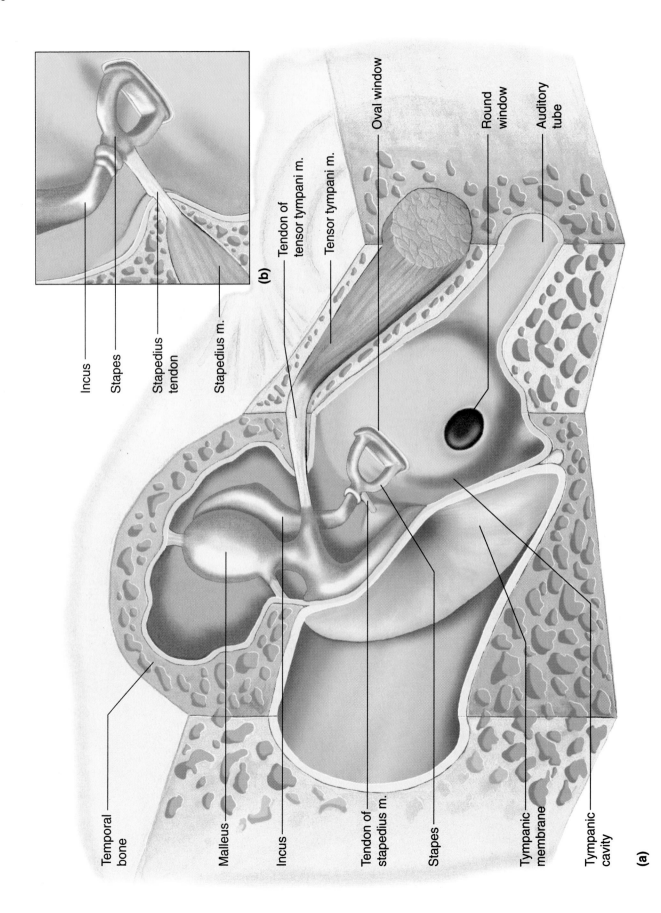

(b)

Incus

Stapes

Stapedius
tendon

Stapedius m.

Tendon of
tensor tympani m.

Tensor tympani m.

Oval window

Round
window

Auditory
tube

Temporal
bone

Malleus

Incus

Tendon of
stapedius m.

Stapes

Tympanic
membrane

Tympanic
cavity

(a)

Figure 12.10   The tensor tympani and the stapedius are effectors in the tympanic reflex

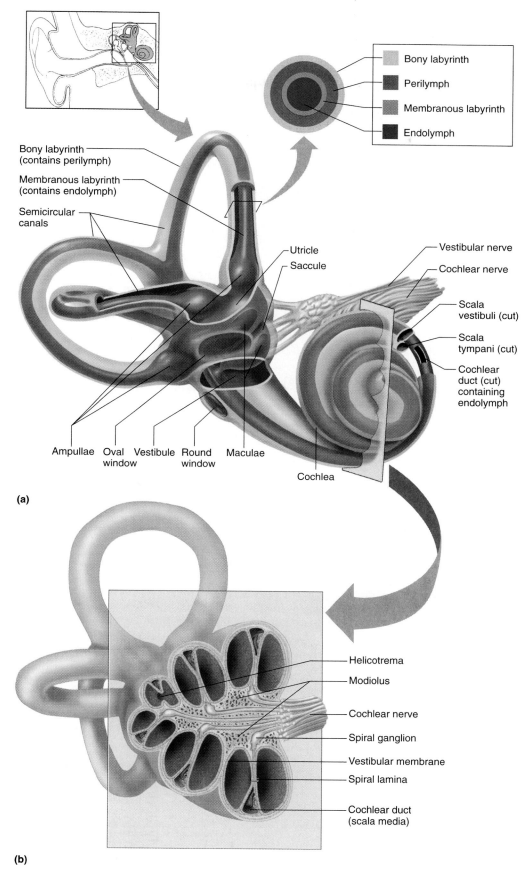

Bony labyrinth

Perilymph

Membranous labyrinth

Endolymph

Bony labyrinth
(contains perilymph)

Membranous labyrinth
(contains endolymph)

Semicircular
canals

Utricle

Saccule

Vestibular nerve

Cochlear nerve

Scala
vestibuli (cut)

Scala
tympani (cut)

Cochlear
duct (cut)
containing
endolymph

Ampullae Oval   Vestibule Round   Maculae
          window   window

Cochlea

(a)

Helicotrema

Modiolus

Cochlear nerve

Spiral ganglion

Vestibular membrane

Spiral lamina

Cochlear duct
(scala media)

(b)

Figure 12.11   (a) Perilymph and (b) the spiral lamina within the inner ear

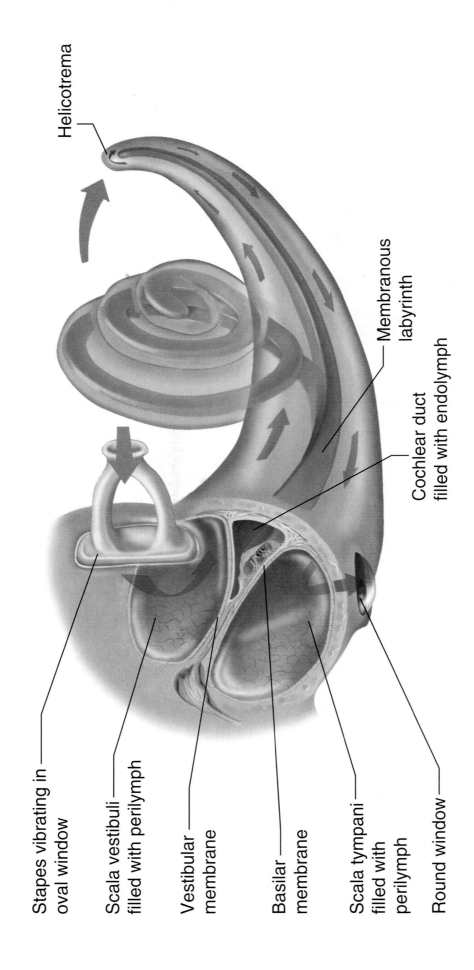

Helicotrema

Membranous
labyrinth

Cochlear duct
filled with endolymph

Stapes vibrating in
oval window

Scala vestibuli
filled with perilymph

Vestibular
membrane

Basilar
membrane

Scala tympani
filled with
perilymph

Round window

Figure 12.12 The cochlea is a coiled, bony canal with a membranous tube (labyrinth) inside

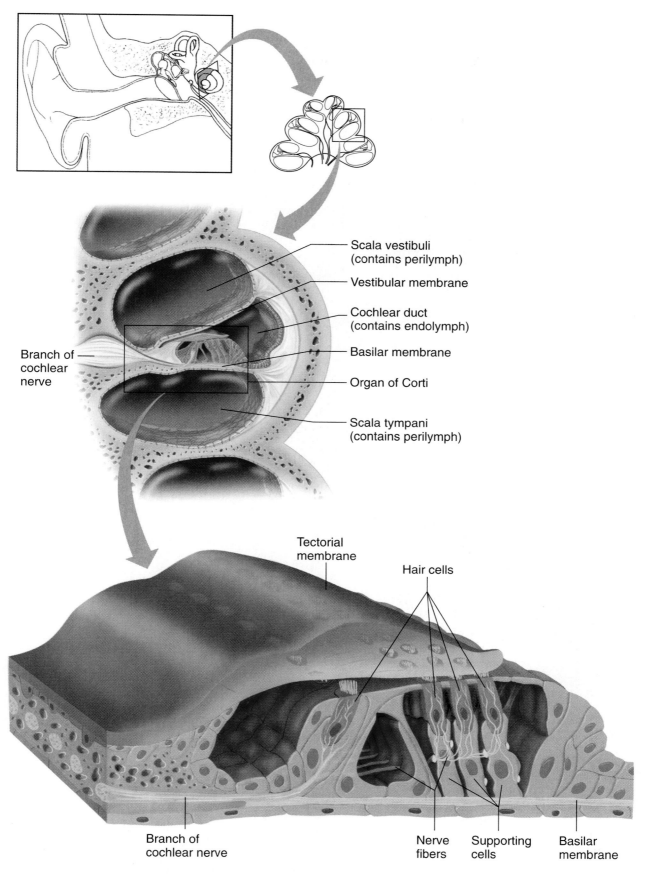

Figure 12.13   Cochlea

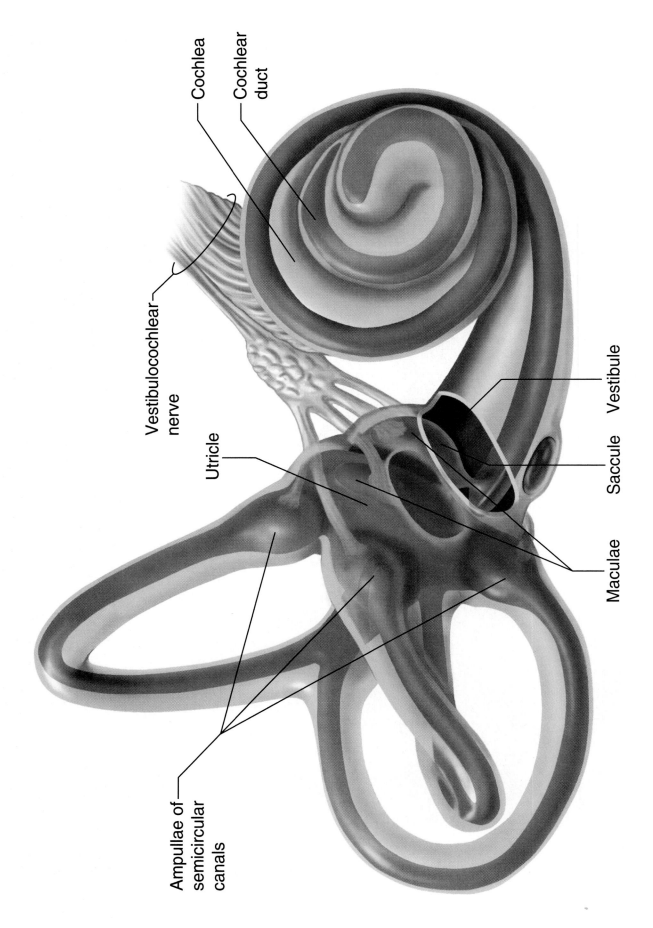

Figure 12.17 The saccule and utricle are located within the bony chamber of the vestibule

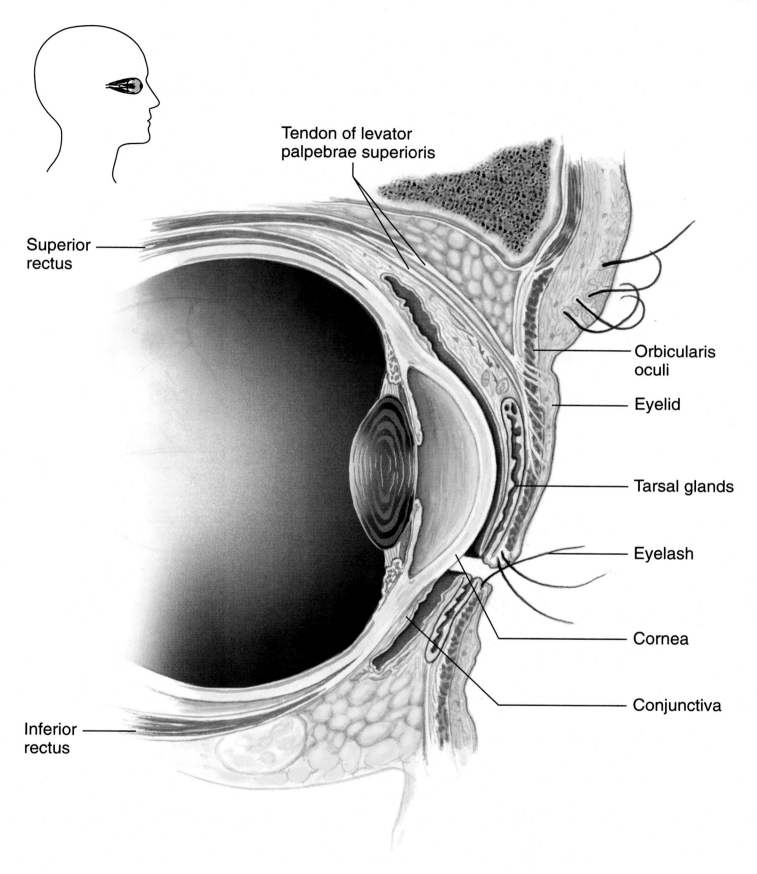

Tendon of levator palpebrae superioris

Superior rectus

Orbicularis oculi

Eyelid

Tarsal glands

Eyelash

Cornea

Conjunctiva

Inferior rectus

Figure 12.22 Saggital section of the closed eyelids and the anterior portion of the eye

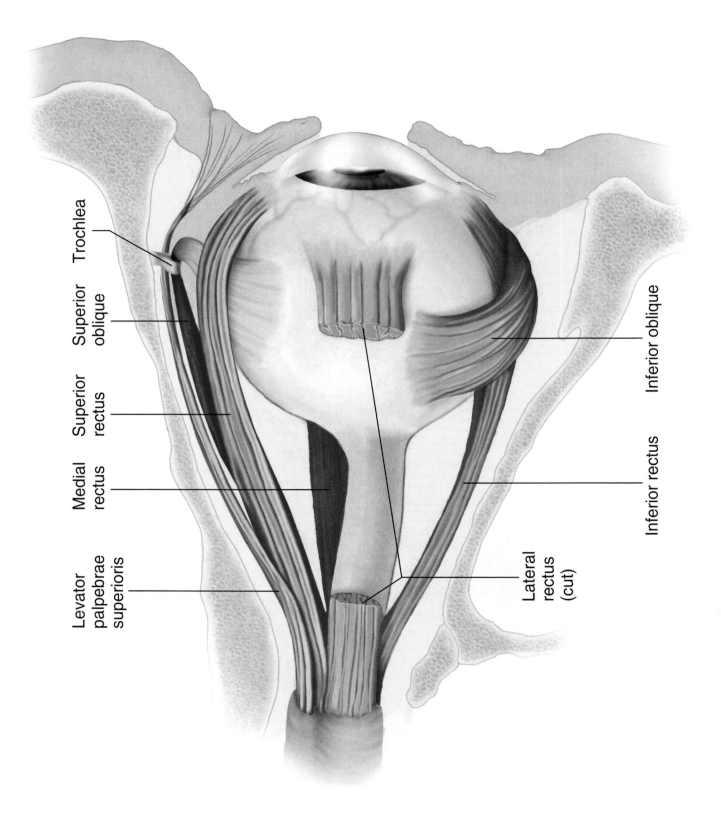

Trochlea

Superior oblique

Superior rectus

Medial rectus

Levator palpebrae superioris

Inferior oblique

Inferior rectus

Lateral rectus (cut)

Figure 12.24   The extrinsic muscles of the right eye (lateral view)

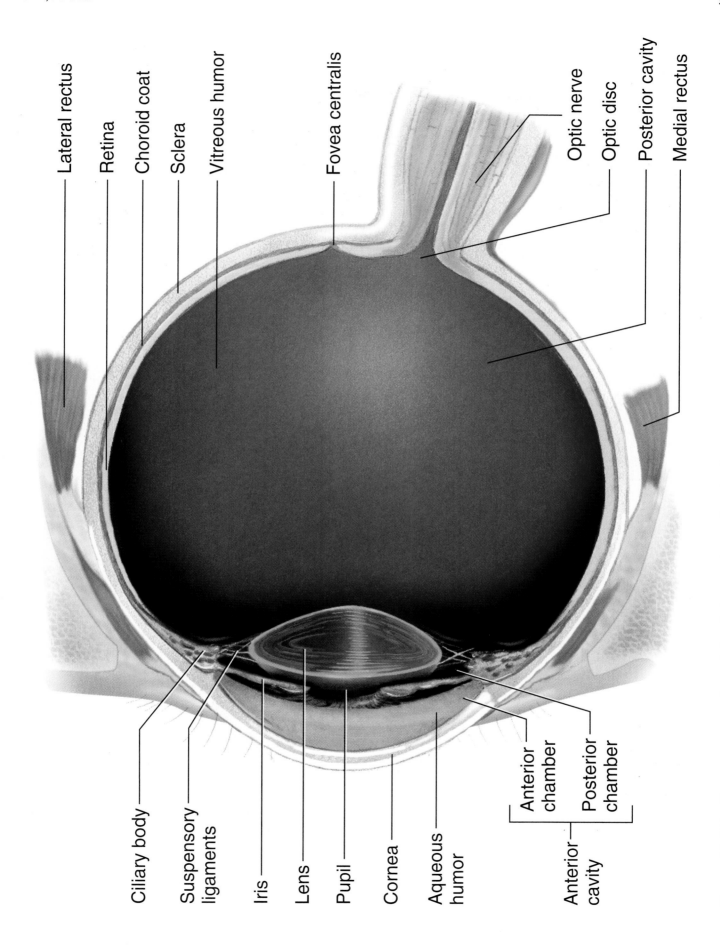

Lateral rectus

Retina

Choroid coat

Sclera

Vitreous humor

Fovea centralis

Optic nerve

Optic disc

Posterior cavity

Medial rectus

Ciliary body

Suspensory ligaments

Iris

Lens

Pupil

Cornea

Aqueous humor

Anterior chamber

Posterior chamber

Anterior cavity

Figure 12.25   Transverse section of the right eye (superior view)

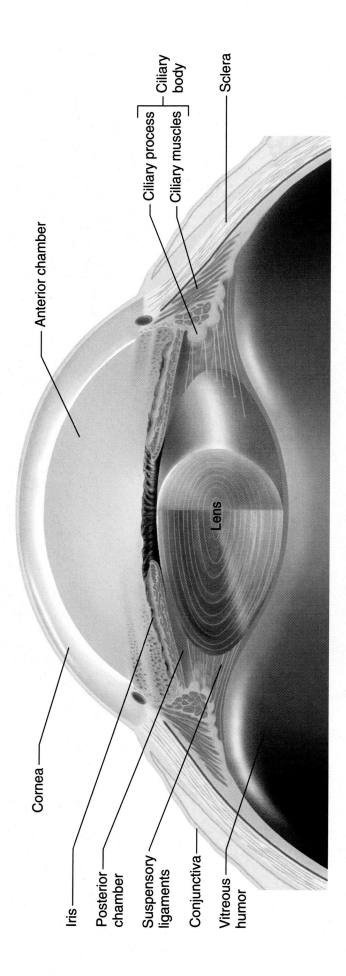

Ciliary process ⎤
             ⎬ Ciliary body
Ciliary muscles ⎦

Sclera

Anterior chamber

Lens

Cornea

Iris

Posterior chamber

Suspensory ligaments

Conjunctiva

Vitreous humor

Figure 12.26   Anterior portion of the eye

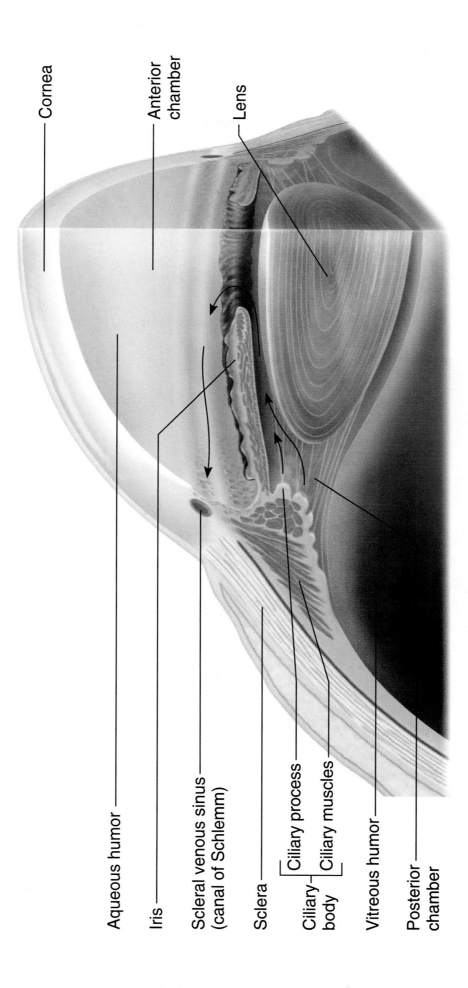

Cornea

Anterior chamber

Lens

Aqueous humor

Iris

Scleral venous sinus (canal of Schlemm)

Sclera

Ciliary body { Ciliary process, Ciliary muscles

Vitreous humor

Posterior chamber

Figure 12.30 Aqueous humor

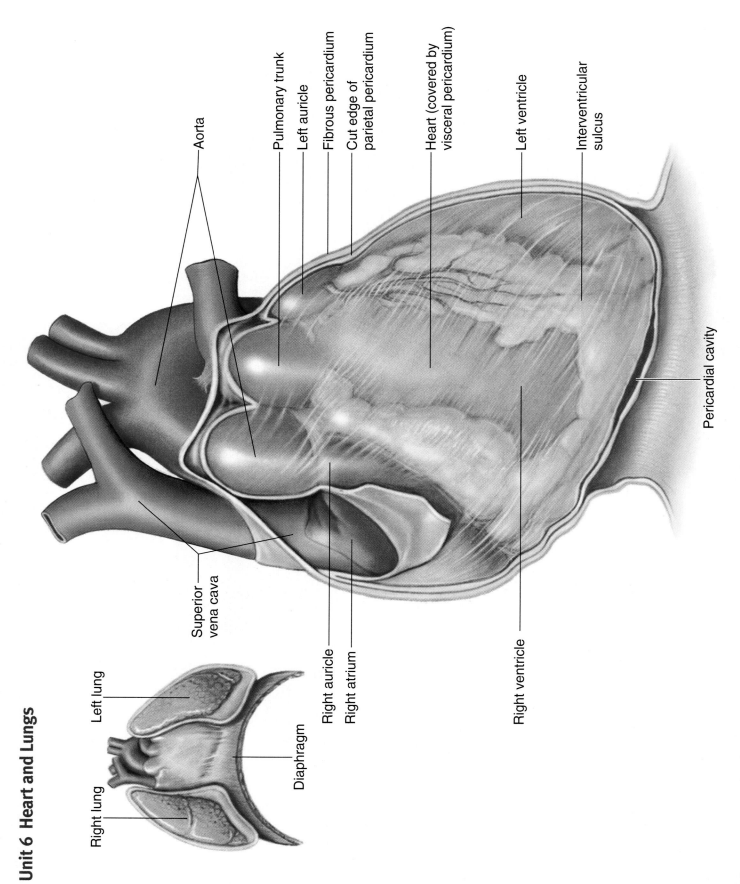

Unit 6 Heart and Lungs

Aorta

Pulmonary trunk

Left auricle

Fibrous pericardium

Cut edge of
parietal pericardium

Heart (covered by
visceral pericardium)

Left ventricle

Interventricular
sulcus

Pericardial cavity

Superior
vena cava

Right auricle

Right atrium

Right ventricle

Left lung

Right lung

Diaphragm

Figure 15.4   The heart is within the mediastinum and is enclosed by a layered pericardium

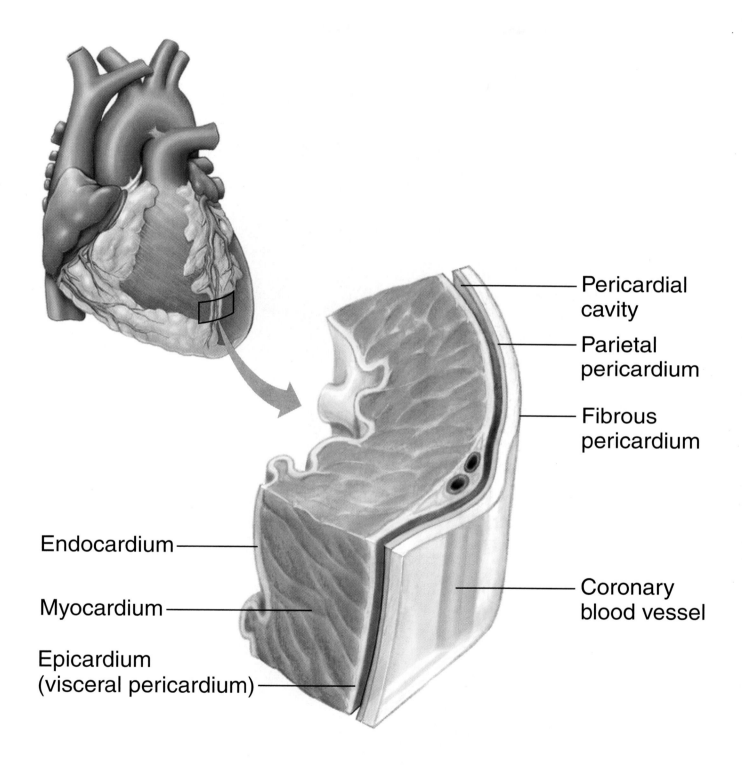

Endocardium

Myocardium

Epicardium
(visceral pericardium)

Pericardial
cavity

Parietal
pericardium

Fibrous
pericardium

Coronary
blood vessel

Figure 15.5 The heart wall has three layers: an endocardium, a myocardium, and an epicardium

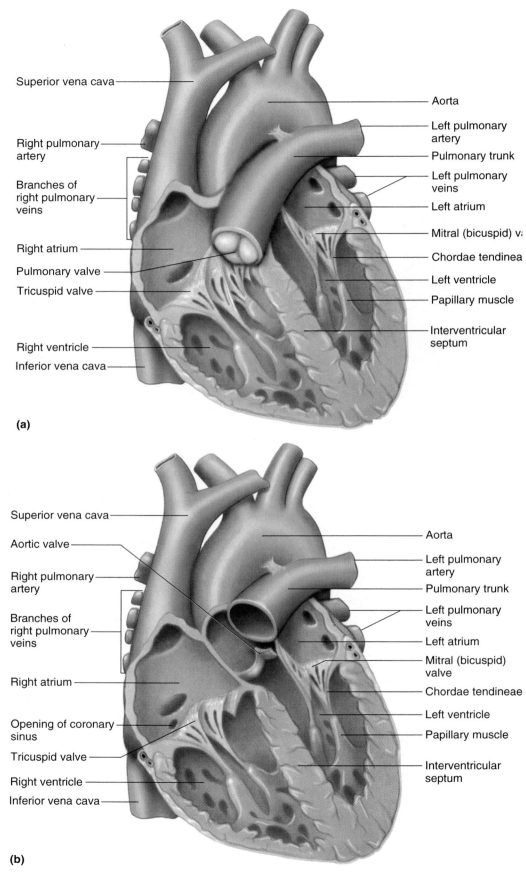

Superior vena cava

Aorta

Right pulmonary
artery

Left pulmonary
artery

Pulmonary trunk

Branches of
right pulmonary
veins

Left pulmonary
veins

Left atrium

Mitral (bicuspid) va

Right atrium

Chordae tendinea

Pulmonary valve

Left ventricle

Tricuspid valve

Papillary muscle

Interventricular
septum

Right ventricle

Inferior vena cava

(a)

Superior vena cava

Aorta

Aortic valve

Left pulmonary
artery

Right pulmonary
artery

Pulmonary trunk

Branches of
right pulmonary
veins

Left pulmonary
veins

Left atrium

Mitral (bicuspid)
valve

Right atrium

Chordae tendineae

Left ventricle

Opening of coronary
sinus

Papillary muscle

Tricuspid valve

Interventricular
septum

Right ventricle

Inferior vena cava

(b)

Figure 15.6   Coronal sections of the heart

Aortic valve

Tricuspid valve

Pulmonary valve

Opening of left coronary artery

Mitral valve

Fibrous skeleton

Posterior

Figure 15.9   The skeleton of the heart (superior view)

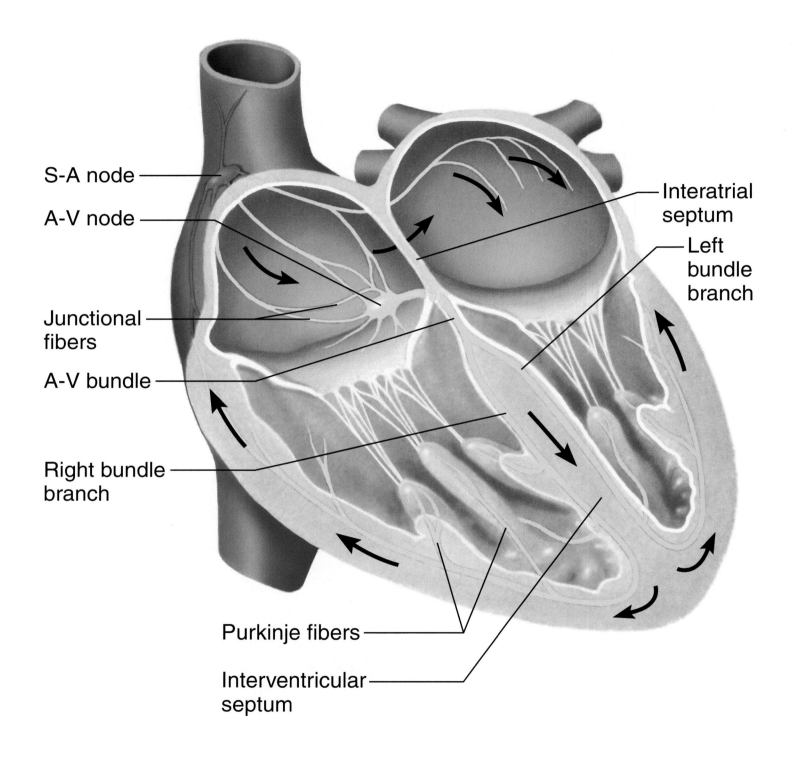

Figure 15.19   The cardiac conduction system

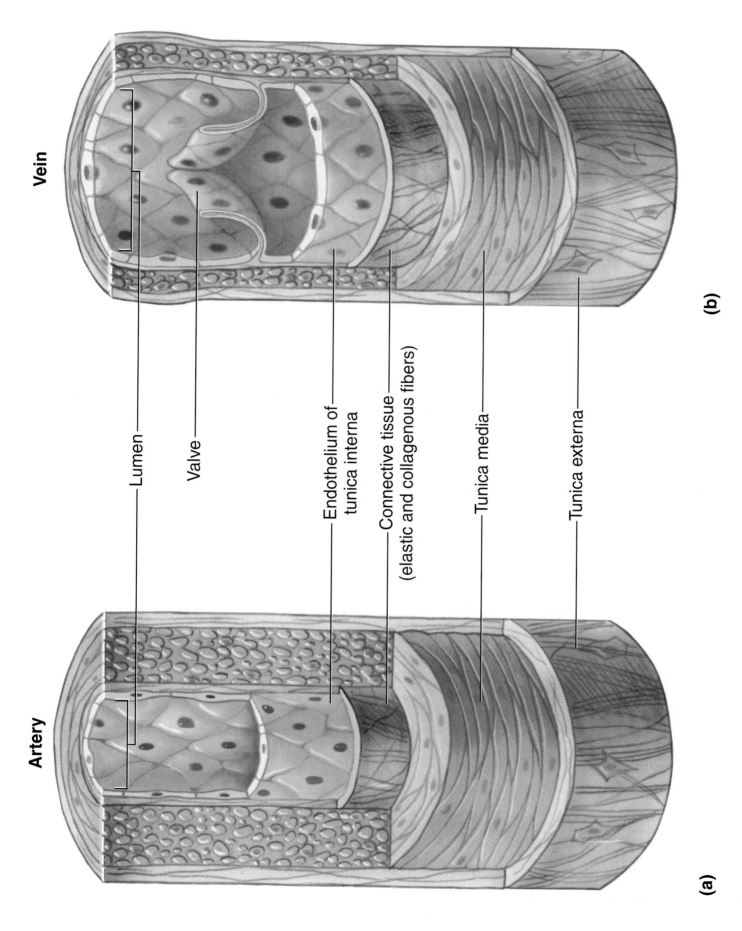

Figure 15.25   Blood vessels

Vein

Artery

Lumen

Valve

Endothelium of tunica interna

Connective tissue (elastic and collagenous fibers)

Tunica media

Tunica externa

(a)

(b)

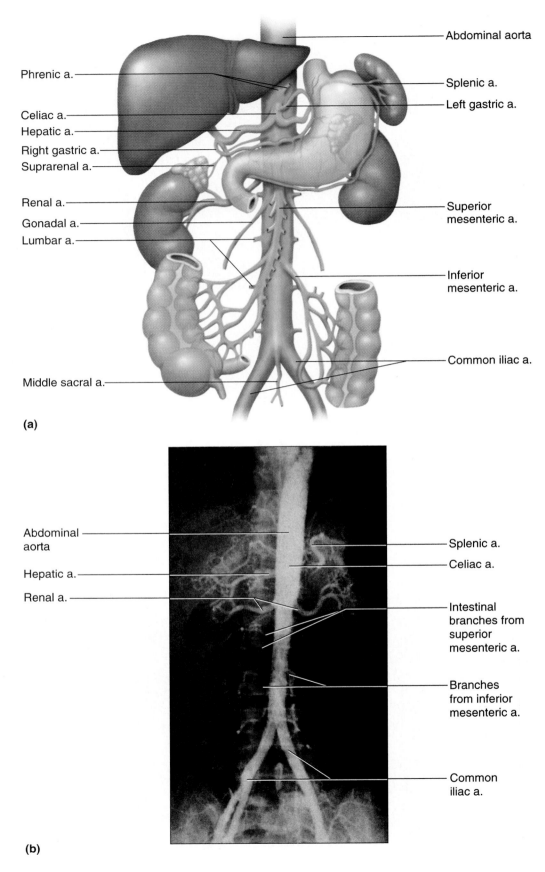

Phrenic a.

Celiac a.
Hepatic a.
Right gastric a.
Suprarenal a.

Renal a.
Gonadal a.
Lumbar a.

Middle sacral a.

(a)

Abdominal aorta

Splenic a.
Left gastric a.

Superior mesenteric a.

Inferior mesenteric a.

Common iliac a.

Abdominal
aorta

Hepatic a.

Renal a.

(b)

Splenic a.
Celiac a.

Intestinal branches from superior mesenteric a.

Branches from inferior mesenteric a.

Common iliac a.

Figure 15.45   Abdominal aorta

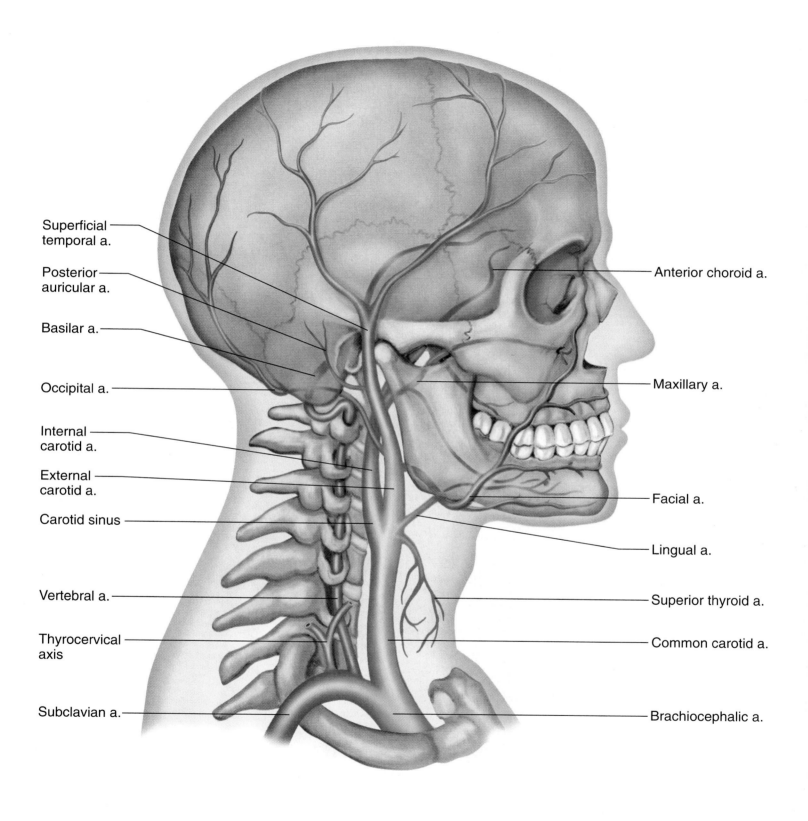

Superficial temporal a.

Posterior auricular a.

Basilar a.

Occipital a.

Internal carotid a.

External carotid a.

Carotid sinus

Vertebral a.

Thyrocervical axis

Subclavian a.

Anterior choroid a.

Maxillary a.

Facial a.

Lingual a.

Superior thyroid a.

Common carotid a.

Brachiocephalic a.

Figure 15.46   The main arteries of the head and neck

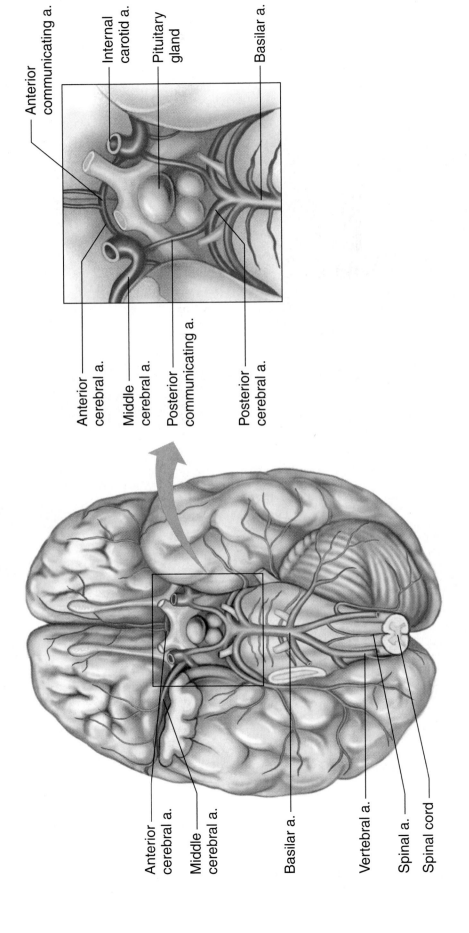

Anterior communicating a.

Internal carotid a.

Pituitary gland

Basilar a.

Anterior cerebral a.

Middle cerebral a.

Posterior communicating a.

Posterior cerebral a.

Anterior cerebral a.

Middle cerebral a.

Basilar a.

Vertebral a.

Spinal a.

Spinal cord

Figure 15.48 **View of inferior surface of the brain**

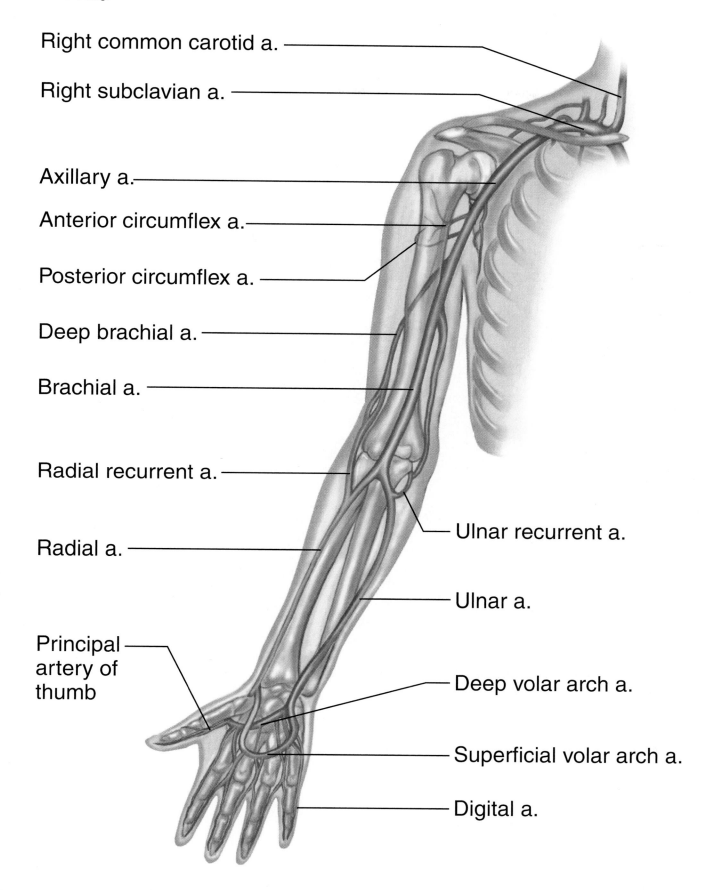

Right common carotid a.

Right subclavian a.

Axillary a.

Anterior circumflex a.

Posterior circumflex a.

Deep brachial a.

Brachial a.

Radial recurrent a.

Radial a.

Principal artery of thumb

Ulnar recurrent a.

Ulnar a.

Deep volar arch a.

Superficial volar arch a.

Digital a.

Figure 15.49 **The main arteries to the shoulder and upper limb**

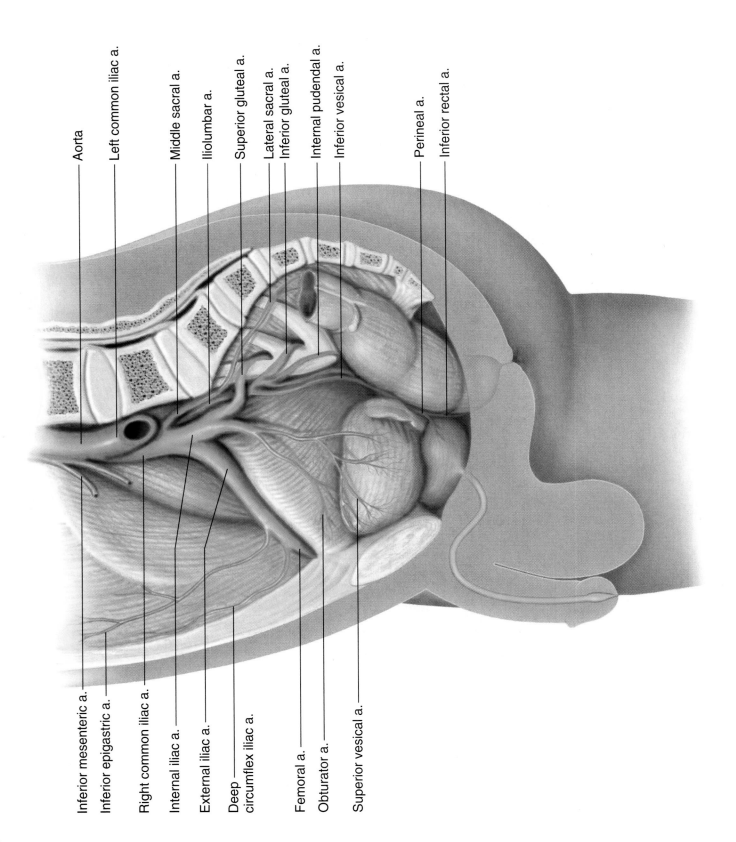

Figure 15.51   Arteries that supply the pelvic region

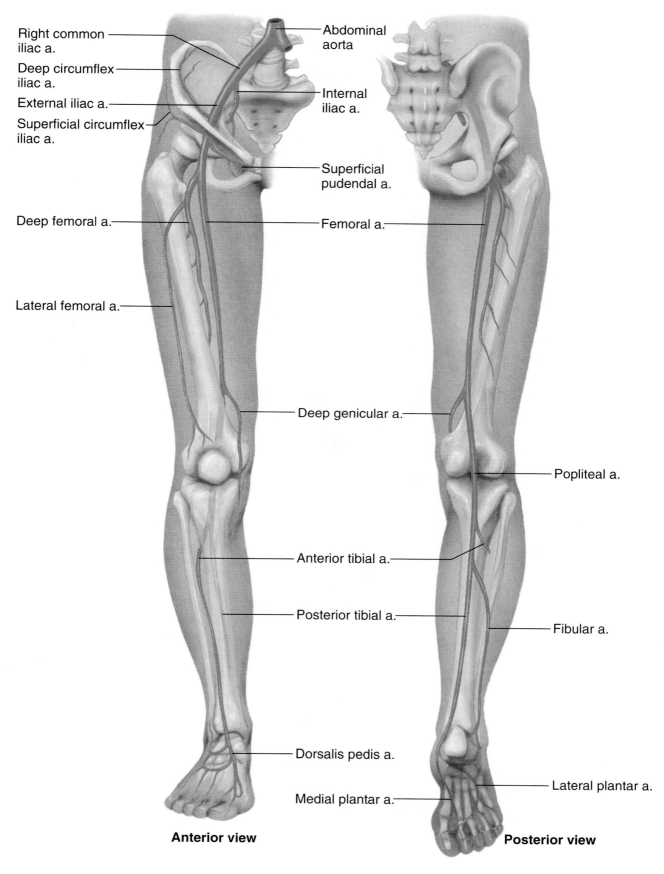

Right common iliac a.

Deep circumflex iliac a.

External iliac a.

Superficial circumflex iliac a.

Abdominal aorta

Internal iliac a.

Superficial pudendal a.

Deep femoral a.

Femoral a.

Lateral femoral a.

Deep genicular a.

Popliteal a.

Anterior tibial a.

Posterior tibial a.

Fibular a.

Dorsalis pedis a.

Lateral plantar a.

Medial plantar a.

Anterior view

Posterior view

Figure 15.52   Main branches of the external iliac artery

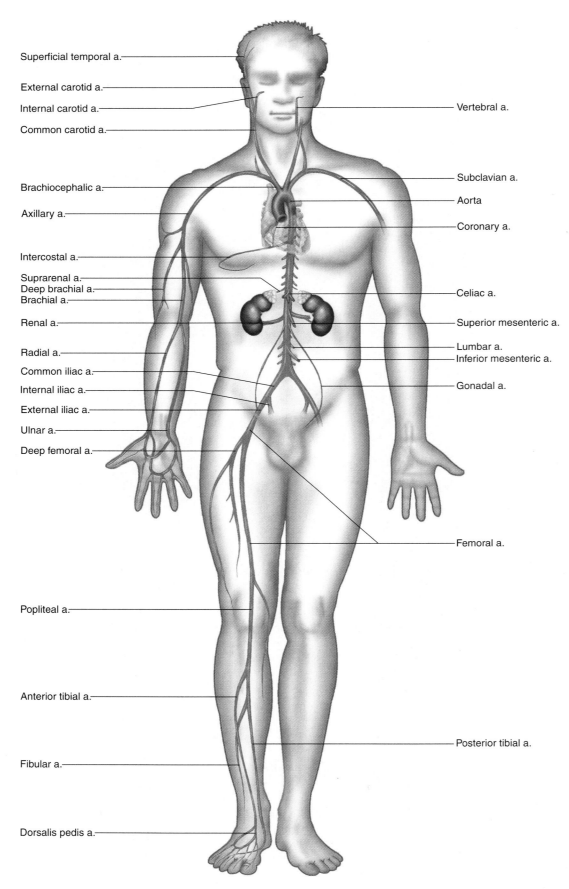

Superficial temporal a.

External carotid a.

Internal carotid a.

Common carotid a.

Vertebral a.

Brachiocephalic a.

Subclavian a.

Aorta

Axillary a.

Coronary a.

Intercostal a.

Suprarenal a.

Deep brachial a.

Brachial a.

Celiac a.

Renal a.

Superior mesenteric a.

Radial a.

Lumbar a.

Inferior mesenteric a.

Common iliac a.

Internal iliac a.

Gonadal a.

External iliac a.

Ulnar a.

Deep femoral a.

Femoral a.

Popliteal a.

Anterior tibial a.

Posterior tibial a.

Fibular a.

Dorsalis pedis a.

Figure 15.53   Major vessels of the arterial system

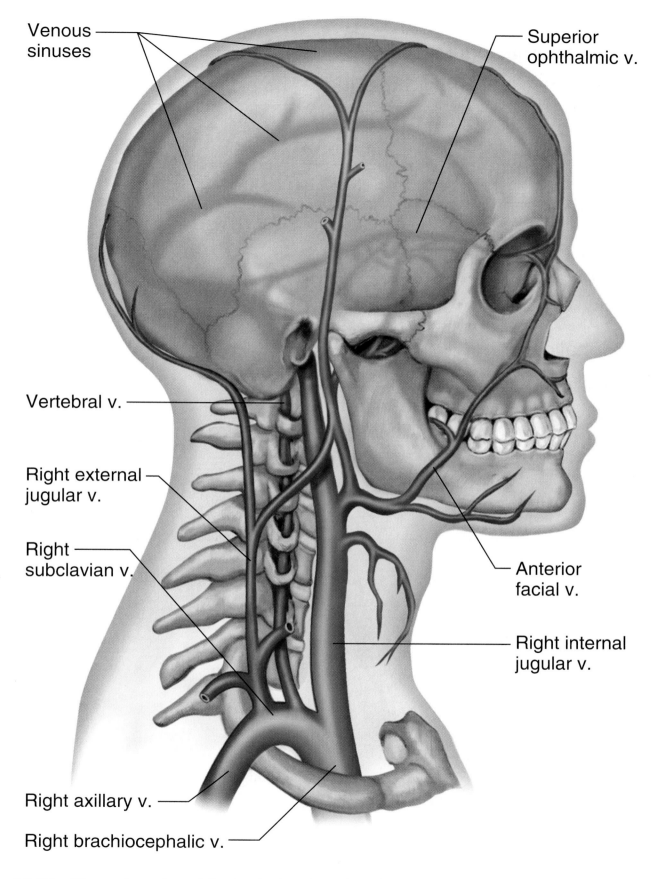

Venous sinuses

Superior ophthalmic v.

Vertebral v.

Right external jugular v.

Right subclavian v.

Anterior facial v.

Right internal jugular v.

Right axillary v.

Right brachiocephalic v.

Figure 15.54 **The major veins of the brain, head, and neck**

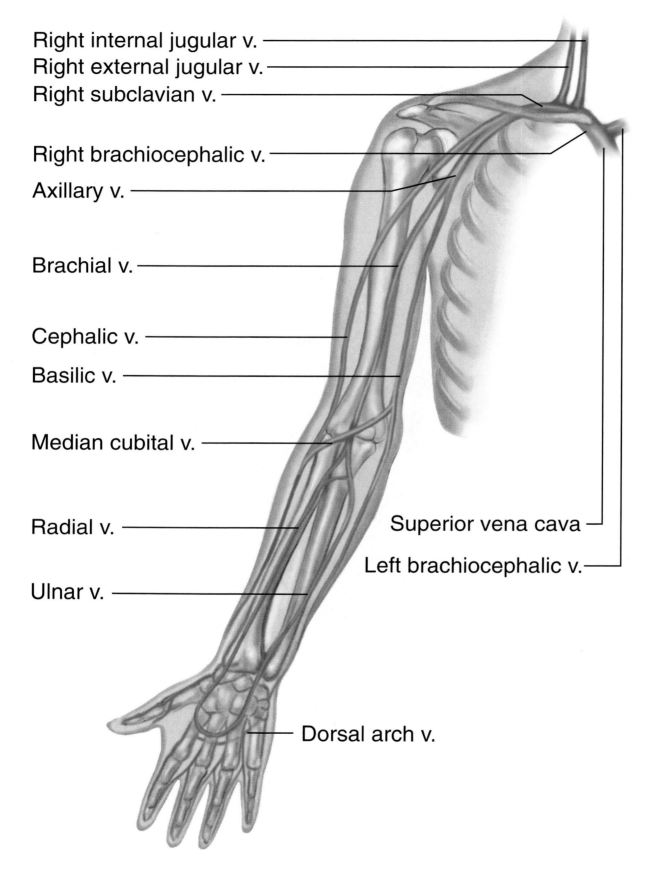

Right internal jugular v.
Right external jugular v.
Right subclavian v.

Right brachiocephalic v.
Axillary v.

Brachial v.

Cephalic v.
Basilic v.

Median cubital v.

Radial v.

Ulnar v.

Superior vena cava
Left brachiocephalic v.

Dorsal arch v.

Figure 15.55   The main veins of the upper limb and shoulder

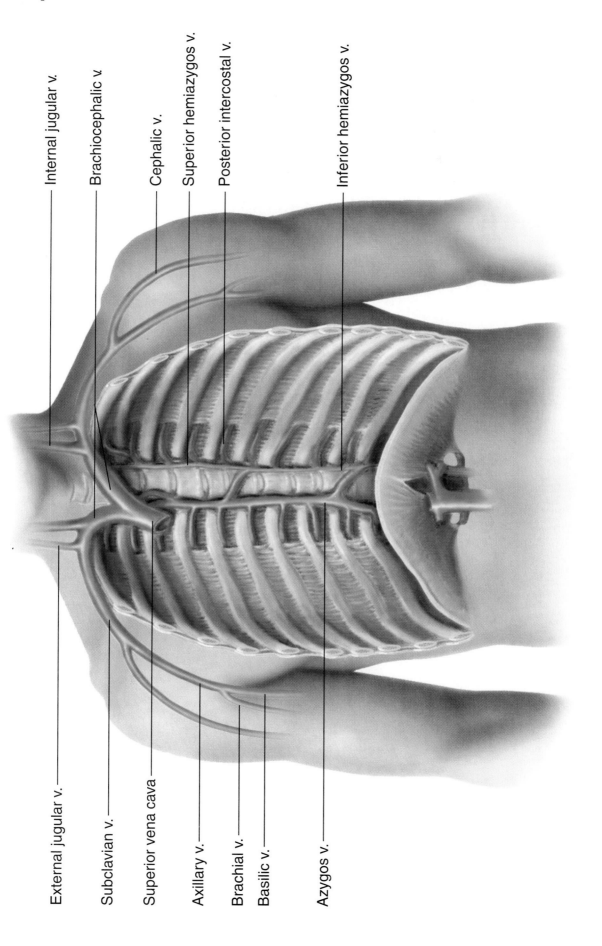

Internal jugular v.

Brachiocephalic v.

Cephalic v.

Superior hemiazygos v.

Posterior intercostal v.

Inferior hemiazygos v.

External jugular v.

Subclavian v.

Superior vena cava

Axillary v.

Brachial v.

Basilic v.

Azygos v.

Figure 15.56 Veins that drain the thoracic wall

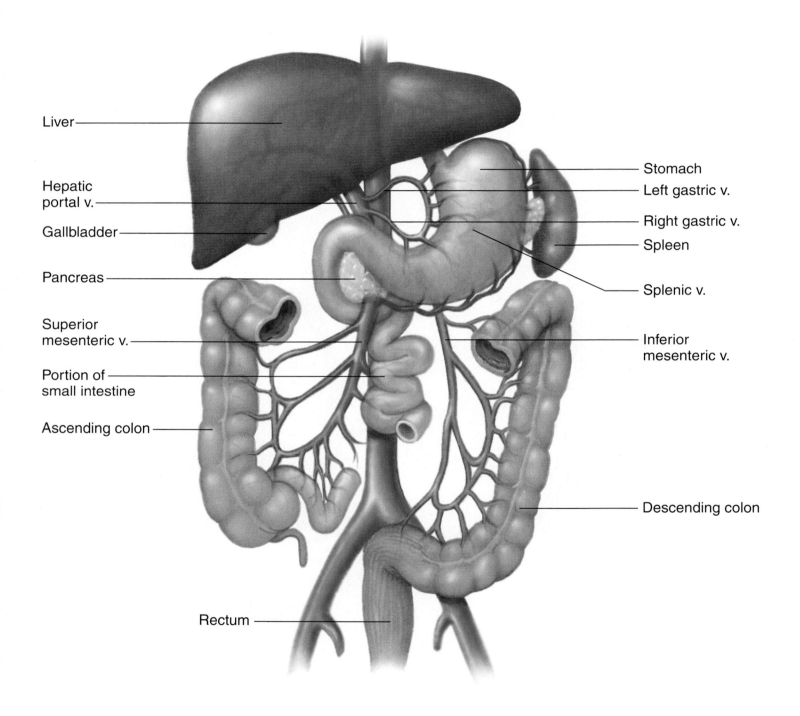

Liver

Hepatic
portal v.

Gallbladder

Pancreas

Superior
mesenteric v.

Portion of
small intestine

Ascending colon

Rectum

Stomach

Left gastric v.

Right gastric v.

Spleen

Splenic v.

Inferior
mesenteric v.

Descending colon

Figure 15.57   Veins that drain the abdominal viscera

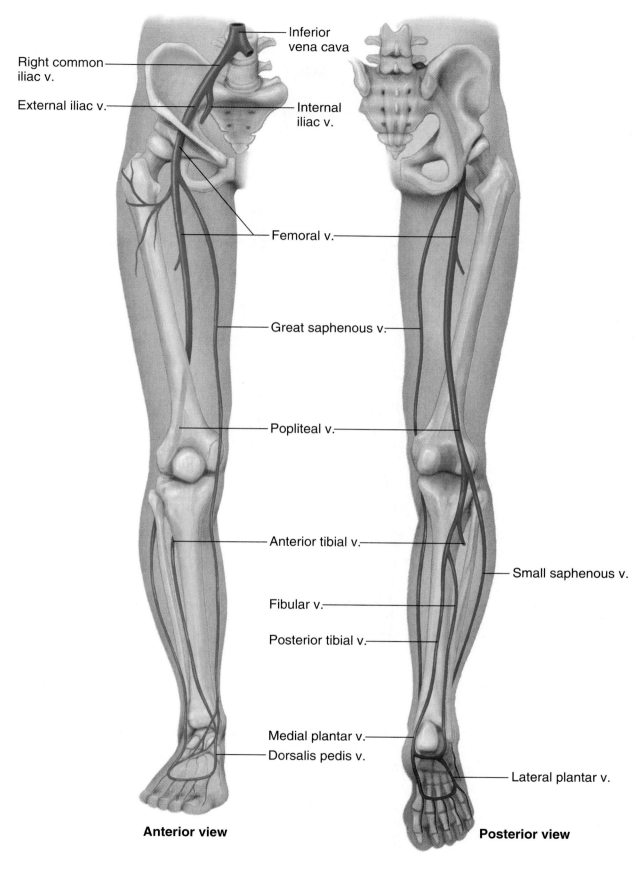

Inferior vena cava

Right common iliac v.

External iliac v.

Internal iliac v.

Femoral v.

Great saphenous v.

Popliteal v.

Anterior tibial v.

Small saphenous v.

Fibular v.

Posterior tibial v.

Medial plantar v.
Dorsalis pedis v.

Lateral plantar v.

Anterior view

Posterior view

Figure 15.59   The main veins of the lower limb and pelvis

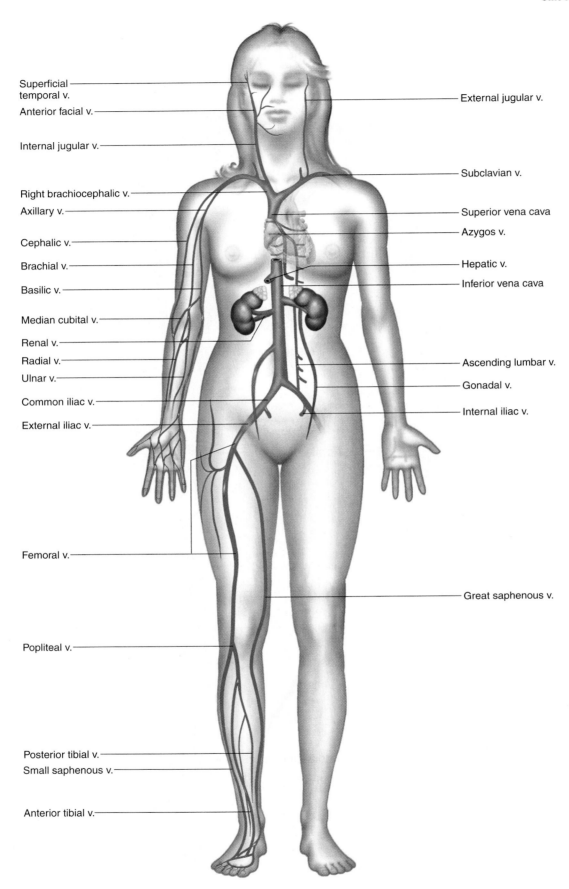

Superficial temporal v.

Anterior facial v.

Internal jugular v.

Right brachiocephalic v.

Axillary v.

Cephalic v.

Brachial v.

Basilic v.

Median cubital v.

Renal v.

Radial v.

Ulnar v.

Common iliac v.

External iliac v.

Femoral v.

Popliteal v.

Posterior tibial v.

Small saphenous v.

Anterior tibial v.

External jugular v.

Subclavian v.

Superior vena cava

Azygos v.

Hepatic v.

Inferior vena cava

Ascending lumbar v.

Gonadal v.

Internal iliac v.

Great saphenous v.

Figure 15.60   **Major vessels of the venous system**

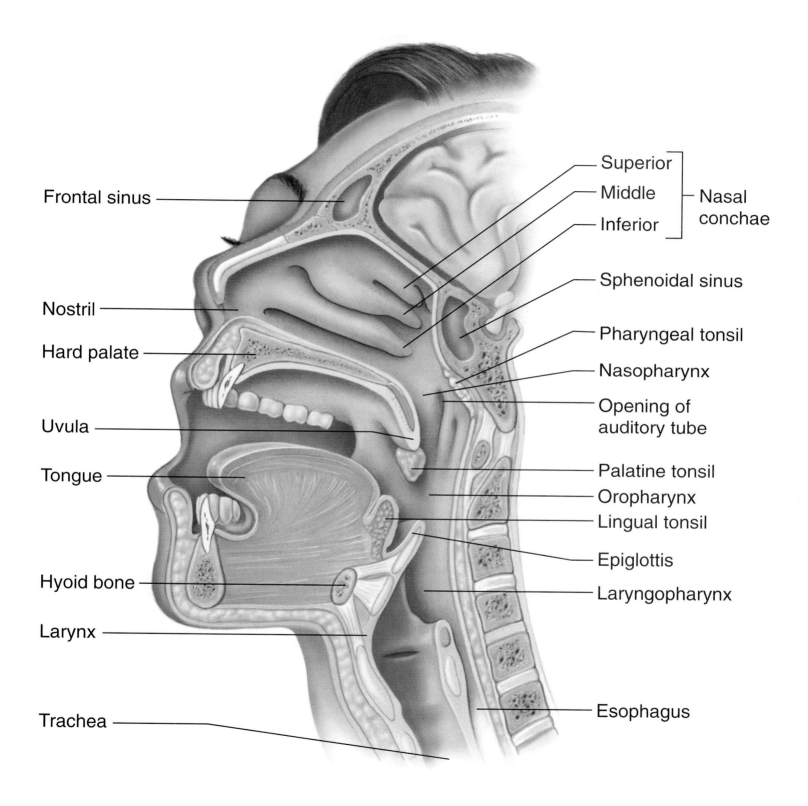

Frontal sinus

Nostril

Hard palate

Uvula

Tongue

Hyoid bone

Larynx

Trachea

Superior
Middle
Inferior
} Nasal conchae

Sphenoidal sinus

Pharyngeal tonsil

Nasopharynx

Opening of auditory tube

Palatine tonsil

Oropharynx

Lingual tonsil

Epiglottis

Laryngopharynx

Esophagus

Figure 19.2   Major features of the upper respiratory tract

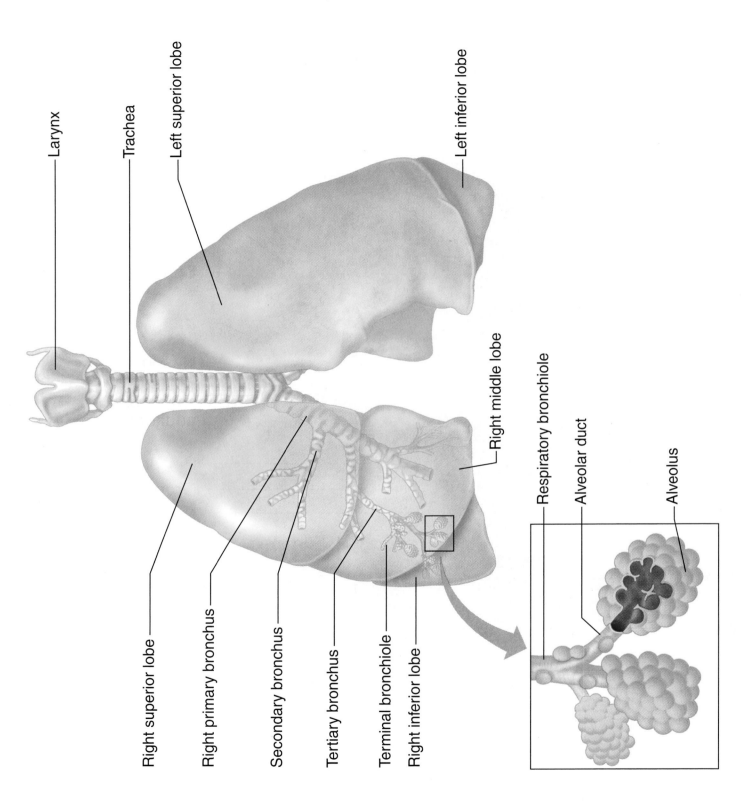

Larynx

Trachea

Left superior lobe

Left inferior lobe

Right superior lobe

Right primary bronchus

Secondary bronchus

Tertiary bronchus

Terminal bronchiole

Right inferior lobe

Right middle lobe

Respiratory bronchiole

Alveolar duct

Alveolus

Figure 19.12   The bronchial tree consists of the passageways that connect the trachea and the alveoli

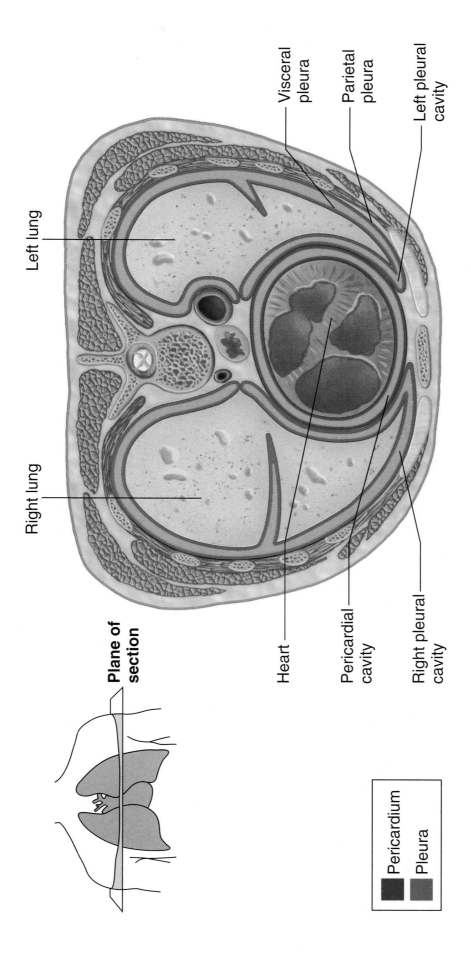

Visceral pleura

Parietal pleura

Left pleural cavity

Left lung

Right lung

Plane of section

Heart

Pericardial cavity

Right pleural cavity

Pericardium

Pleura

Figure 19.20   The left and right pleural cavities